D0139828

THE DIRECTION OF TIME

HANS REICHENBACH

THE DIRECTION OF TIME

EDITED BY MARIA REICHENBACH

DOVER PUBLICATIONS, INC.
Mineola, New York

Bibliographical Note

This Dover edition, first published in 1999, is an unabridged republication of the paperback edition (1971) of the work originally published by the University of California Press, Berkeley, in 1956.

Library of Congress Cataloging-in-Publication Data

Reichenbach, Hans, 1891–1953.
    The direction of time / Hans Reichenbach ; edited by Maria Reichenbach.
        p. cm.
    Originally published: Berkeley : University of California Press, 1956.
    Includes index.
    ISBN 0-486-40926-0 (pbk.)
    1. Causality (Physics) 2. Space and time. 3. Science—Philosophy. I. Reichenbach, Maria. II. Title.

QC6.4.C3 R45 1999
530.11—dc21

                                                                99-049141

Manufactured in the United States of America
Dover Publications, Inc., 31 East 2nd Street, Mineola, N.Y. 11501

• Preface

Among the various manuscripts which my husband left at the time of his sudden death in April, 1953, was a major work on "The Direction of Time", which is published posthumously in this book.

The subject of time has occupied the minds of philosophers of all ages. The first chapter of the present book is in part a historical survey and critical evaluation of the different interpretations offered by traditional philosophical systems. Moreover, the topic of time is very much at the center of recent discussions, among physicists as well as among philosophers. On the basis of the results of modern science, however, it has acquired a character very different from that of traditional philosophy. Since scientific philosophy emphasizes the indispensable connection between the sciences and the philosophical explication of their underlying concepts, the rapid changes in the conceptions of the physicist will naturally influence and guide the philosopher of science. This book gives such an explication of some of the properties of physical time, taking into account the pertinent evidence from physics now available.

My husband's interest in the analysis of time can be traced to his earliest writings. In his main works on the theory of relativity, for instance in *Relativitaetstheorie und Erkenntnis apriori* and in his *Philosophie der Raum-Zeit-Lehre*, he dealt especially with the quantitative, or metrical, properties of time, whereas in the present book he discusses its qualitative, or topological, attributes, such as order and direction. He was convinced that he had achieved a solution of the problem of the direction of time which answers all the questions that can reasonably be asked about it. However, many questions which other philosophers would like to dismiss as pseudo-problems—for instance, the question why we cannot influence the past or the question of the freedom of will—were regarded by him as genuine.

The plan of the book had included a final chapter which, unfortunately, my husband did not live to write. Yet, in spite of the fact that the last chapter is missing, the analysis of *physical* time is complete; the book is therefore not to be considered a fragment. The last chapter would have dealt with the problem of the connection between the subjective experience of time by a human organism and the objective properties of time in nature. In this chapter on the human mind, the qualitative properties of time discussed in §2 would have been reconsidered on the basis of the analysis of physical time in §§3–30. The fact that the direction of subjective time coincides with the positive direction of physical time was to be explained by the conception of human memory as a natural registering instrument that keeps records of the past similar to other natural or artificial records. The last chapter would therefore have constituted an application of the general theory of physical time to a special case, rather than an elaboration of that theory. An appendix to this book contains a brief statement of some of the ideas which would have been developed in the final chapter.

Shortly before his death, my husband received an invitation from Harvard University to give the William James Lectures in the fall semester of 1953. He chose "Time and Free Will" as the title of these lectures and intended to base them on the manuscript of the present book, omitting its more technical parts in order to reach a wider audience.

A kind of summary of the results of my husband's research concerning the direction of time was given in a series of lectures delivered in June, 1952, at the Institut Henri Poincaré, of the Sorbonne, in Paris. These lectures were addressed to physicists and mathematicians and centered on quantum mechanics. They have since been published.[1]

Upon this last book of his, my husband brought to bear all the results of his former philosophical investigations, and he often looked upon this work as a culmination and integration of his main philosophical inquiries and contributions. He was thinking, for instance, of the results of his analysis of probability, quantum mechanics, the theory of relativity, and causality, which play an important part in this book. Furthermore, it is interesting to note that his theory of equivalent descriptions finds a number of new applications here. And the book exemplifies the analytic methods of scientific philosophy which he not only applied but also helped to develop and refine.

It may well be that he would have made slight revisions in the text before it went to the publisher. As a matter of fact, a small amount of editing has been necessary, but no substantial changes have been made. A few footnotes have been added, which are clearly distinguished from the original footnotes by my initials and by inclusion in brackets.

For their detailed advice and generous assistance in the final preparation of the manuscript for publication, I should like to express my sincere appreciation and gratitude to Professor Alfred Landé, Professor Adolf Grünbaum, Dr. Wesley Salmon, Dr. Norman Dalkey, Dr. Albert Latter, Mrs. Ruth Anna Mathers, and Mr. Robert L. Mathers.

Mr. James Murray and Mr. Thomas Stantial drafted the diagrams for the book.

I am also indebted to Miss Helen Travis for her patience in editing the manuscript and to the University of California Press for doing all in its power to facilitate the publication of this book.

MARIA REICHENBACH

Los Angeles, May, 1955

---

[1] "Les Fondements logiques de la mécanique des quanta", *Annales de l'Institut Henri Poincaré*, Tome XIII (Paris, 1953), Fasc. II, pp. 109-158.

• *Contents*

I. Introduction

II. The Time Order of Mechanics

V. The Time of Quantum Physics

CHAPTER | INTRODUCTION

● 1. The Emotive Significance of Time

The problem of time has always baffled the human mind. Not only the events of the external world but even all our subjective experiences occur in time. It appears as though the flow of time, which orders the events of the physical world, passes through human consciousness and compels it to adjust itself to the same order. Our observations of physical things, our feelings and emotions, and our thinking processes extend through time and cannot escape the steady current that flows unhaltingly from the past by way of the present to the future.

What we experience in one moment, glides, in the next moment, into the past. There it remains forever, irretrievable, exempt from further change, inaccessible to further control by anything that the future will bring us—and yet enshrined in our memory as something that once filled our experience as an immediate present. Will it never come back? Why can it not be with us a second time?

1

Undisturbed by our query, the flow of time goes on. Already our present is filled by other experiences which, at the earlier time, we could not completely anticipate. Though in part predictable, the present experience contains many unexpected and previously unknowable features. What was uncertain is now determined. Possibilities which we feared, or hoped for, have now become realities; others, which we never had thought of, have intervened. And even the familiar daily experiences, though highly predictable, reveal in their actual occurrences some specific characteristics that could not have been foreseen. What else awaits us in the future? Will there be a war, or some other political catastrophe? Shall we get the long-hoped-for salary raise? Will a letter arrive that tells us about the death of a friend whom we believed to be in good health? Or will a letter announce that some distant relative bequeathed a fortune to us? And what will the little things which we expect be like? Will the car start right away? Shall we get through the intersection before the traffic light turns red? What will Fred say when I tell him that Doris is going to marry John?

All these things are in the future. What is it, this future? Does it keep events in stock, so to speak, and distribute them according to a plan? Or do events grow from chance? Growing means becoming. What is Becoming? How can something unreal become real? And as soon as it is real, it slides into the past, only to become unreal again, leaving nothing but a shadow in our memory. The present is the only reality. While it slips away, we enter into a new present, thus always remaining in the eternal Now. What is time, if all we have of it is this Now, this one moment gliding with us through the current of events that flows from the unchangeable past to the unknowable future?

Questions of this kind reveal the highly emotional content associated with the experience of time. They tempt us to look for answers that satisfy emotions rather than clarify meanings. I do not wish to say that such questions are unreasonable. But the answers to them may look very different from what we expect; and we may even be unable to find the answers, unless we first revise the questions and make precise what, at this stage, is mere groping for meanings. Human thought processes do not follow the pattern of calculating machines, which have an answer to any question, provided the question is asked correctly. We cannot answer every correct question—but we can often answer questions which are not correctly asked, by first giving them a form in which they have meaning. Often the process of reformulating the question and giving the answer is the same process. Looking for answers, we discover new meanings and find out what it was that we were asking for.

This is the scientific approach. Do not expect answers before you have found clear meanings. But do not throw away unclear questions. Keep them on file until you have the means at the same time to clarify and to answer them. Often these means result from developments in other fields, which at first sight appear to have nothing to do with the question.

The history of philosophy offers many illustrations of this process of clarification of meanings. Thales of Miletus believed that water is the substance of which all things are made. Heraclitus argued that, instead, this mysterious substance was fire. But neither of them knew precisely what it means to say that a piece of matter is composed of several substances. Modern chemistry has made this meaning precise by its methods of chemical analysis and has shown that neither water nor fire is a chemical element. Another illustration is found in Plato's philosophy. Plato believed that geometrical relations are known through visions of ideas, a reminiscence of experiences which our souls had in a world beyond the heavens long before their terrestrial lives began. Modern mathematics has shown that the act of visualizing geometrical figures can be understood in a this-worldly way: it is a recollection of everyday experiences with objects of our environment. It is the meaning of the term "visualization" that was clarified in this answer to a question. And only with the modern answer did the question assume a distinct meaning.

The inquiry into the nature of time has a similar history. It greatly puzzled the ancients, remained unsolved for two thousand years, and found an answer in developments of modern physics which were not directly concerned with the problem of time, but with that of causality. Before turning to these developments, it may be appropriate to examine more closely the conception of time contained within older philosophical systems, since they reveal the emotional reactions and formulate the logical puzzles which every one of us encounters in the experience of time.

Our emotional response to the flow of time is largely determined by the irresistibility of its passing away. The flow of time is not under our control. We cannot stop it; we cannot turn it back; we have the feeling of being carried away by it, helplessly, like a piece of lumber in the current of a river. We can know the past, but we cannot change it. Our activity can be directed toward the future only. But the future is incompletely known, and unexpected events may turn up which make our plans break down. It is true, the future may also have favorable turns in store. Yet we know that they are limited in number and that adjusting ourselves to what the future may bring cannot help us too much—there is only a limited stretch of time ahead of us, and the end of all this striving and responding to new situations is death. The coming of death is the inescapable result of the irreversible flow of time. If we could stop time, we could escape death—the fact that we cannot makes us ultimately impotent, makes us equals of the piece of lumber drifting in the river current. The fear of death is thus transformed into a fear of time, the flow of time appearing as the expression of superhuman forces from which there is no escape. The phrase "passing away", by means of which we evasively speak of death without using its name, reveals our emotional identification of time flow with death.

Dissatisfied emotion has frequently been projected into logic. In theories of the universe it often reappears in the guise of logical queries and pseudo-logical constructions. A philosopher argues that he has discovered a puzzle of Being which logic cannot solve—he might as well say that he has discovered a fact that arouses his emotional resistance. The fear of death has greatly influenced the logical analysis which philosophers have given of the problem of time. The belief that they had discovered paradoxes in the flow of time is called a "projection" in modern psychological terminology. It functions as a defense mechanism; the paradoxes are intended to discredit physical laws that have aroused deeply rooted emotional antagonism.

Religious philosophers have maintained that the happenings in time do not constitute the sum total of reality. They insist that there is another reality, a higher reality, which is exempt from time flow. Only the inferior reality of human experience is bound to time. The assumed superior reality, strangely enough, has been called *eternal*, which is a term referring to time. But in the language of these philosophers the term no longer pertains to permanent duration, but rather to something existing beyond time, not subject to time flow. Its opposite in the terminology of the church is *secular*, a term originally referring to the time span of a human life (in an extended meaning, of a century), but having assumed the meaning of something subject to time flow and thus something earthly, displaying the inferior nature of physical reality. The desire to survive death and to live eternally, in the sense of an unlimited time, a desire obviously incompatible with physical facts, has thus led to a conception in which eternal life is not life in time, but in a different reality. In order to escape the "passing away" with time, a timeless reality was invented.

Among the ancients, Parmenides and Plato developed such concepts of reality, though in different forms. Parmenides tells us that the higher reality does not come into being and does not pass out of being. "It is uncreated and indestructible; for it is complete, immovable, and without end. Nor was it ever, nor will it be; for now it *is*, all at once, a continuous *one*." And Plato explains that "time is the moving image of eternity". Here "eternity" does not mean "infinite time". It is supposed to denote a reality not controlled by time flow, which, however, is reflected, so to speak, in the river of time. The happenings in time are, at best, an inferior form of reality; for Parmenides, it seems, they are not real at all, but illusions.

Such philosophies are documents of emotional dissatisfaction. They make use of metaphors invented to appease the desire to escape the flow of time and to allay the fear of death. They cannot be brought into a logically consistent form. Yet, strangely enough, they are often presented as the results of logical analysis. The grounds offered for them are the alleged paradoxes of Becoming. Parmenides argues that if there were Becoming, a thing must grow from nothing into something, which he regards as logically impossible. And his successor in the Eleatic school, Zeno, has supplied us with a number of famous paradoxes which, he thought, demonstrate the impossibility of motion and the truth of Parmenides' conception of Being as timeless.

Zeno's paradoxes of motion have often been discussed. He argues that if motion is travel from one point to another, a flying arrow cannot move as long as it is at exactly one point. But how then can it get to the next point? Does it jump through a timeless interval? Obviously not. Therefore motion is impossible. Or consider a race between Achilles and a tortoise, in which the tortoise is given a head start. First Achilles has to reach the point where the tortoise started; but by then, the tortoise has moved to a farther point. Then Achilles has to reach that other point, by which time the tortoise again has reached a farther point; and so on, ad infinitum. Achilles would have to traverse an infinite number of nonzero distances before he could catch up with the tortoise; this he cannot do, and therefore he cannot overtake the tortoise.

Concerning the arrow paradox, we answer today that rest at one point and motion at one point can be distinguished. "Motion" is defined, more precisely speaking, as "travel from one point to another in a finite and nonvanishing stretch of time"; likewise, "rest" is defined as "absence of travel from one point to another in a finite and nonvanishing stretch of time". The term "rest at one point at one moment" is not defined by the preceding definitions. In order to define it, we define "velocity" by a limiting process of the kind used for a differential quotient; then "rest at one point" is defined as the value zero of the velocity. This logical procedure leads to the conclusion that the flying arrow, at each point, possesses a velocity greater than zero and therefore is not at rest. Furthermore, it is not permissible to ask how the arrow can get to the next point, because in a continuum there is no next point. Whereas for every integer there exists a next integer, it is different with a continuum of points: between any two points there is another point. Concerning the other paradox, we argue that Achilles can catch up with the tortoise because an infinite number of nonvanishing distances converging to zero can have a finite sum and can be traversed in a finite time.

These answers, in order to be given in all detail, require a theory of infinity and of limiting processes which was not elaborated until the nineteenth century.[1] In the history of logic and mathematics, therefore, Zeno's paradoxes occupy an important place; they have drawn attention to the fact that the logical theory of the ordered totality of points on a line—the continuum—cannot be given unless the assumption of certain simple regularities displayed by the series of integers is abandoned. In the course of such investigations, mathematicians have discovered that the concept of infinity is capable of a logically consistent treatment, that the infinity of points on a line differs from that of the integers, and that Zeno's paradoxes are not restricted to temporal flow, since they can likewise be formulated and solved for a purely spatial continuum.

What makes Zeno's paradoxes psychologically interesting, however, is the fact that they were discovered, not as part of the pursuit of a mathematical theory of the continuum, but through a process of rationalization; that they were found because the Eleatic school wanted to prove the unreality of time. Had Zeno not constructed his paradoxes under the spell of this preoccupation with a "metaphysical" aim, he would have come to a different solution. He would have argued that, since arrows do fly and a fast runner does overtake a tortoise, there must be something wrong with his conception of logic, but not with physical reality. But he did not want to come to this conclusion. He wanted to show that change and Becoming are illusory, and he wanted to show that Reality has a timeless existence exempt from the shortcomings of time-controlled human experience—from passing away and from death.

The time theory of Parmenides has become the historical symbol of a negative emotional attitude toward the flow of time. But the actual structure of time is compatible with different emotive reactions; and there has always existed a positive attitude toward time flow, an

---

[1]For a modern discussion of Zeno's paradoxes, see Bertrand Russell, *Our Knowledge of the External World* (Chicago, Open Court Publ. Co., 1914, chap. vi; and Adolf Grünbaum, "Relativity and the Atomicity of Becoming", *Rev. Metaphys.*, Vol. 4 (1950), p. 176.

affirmative emotional response to change and Becoming, for which the future is an inexhaustible source of new experiences and a challenge to our abilities to make the best of emerging opportunities. The historical symbol of this positive emotional attitude toward time flow was created in the philosophy of Parmenides' contemporary and opponent Heraclitus.

"All things are in flux" is the formula in which Heraclitus' philosophy has been summed up. Becoming is for him the very essence of life. "The sun is new every day"—this means, for him, that it is good that every day produces something new. We need not cling to what has been; we can get along very well in a world of continuous change. "You cannot step twice into the same river, for fresh waters are ever flowing in upon you." This seeming paradox is not as profound as Heraclitus believed, for we can very well call the river *the same* even though its waters change. But Heraclitus' aphorism draws our attention to the logical nature of the physical thing as a series of different states in time; it is not necessary for physical identity that these states be exactly alike. A human being is the same, identical person all the time, although the body grows and changes its chemical building blocks. A physics of things does not require a denial of time flow. Common sense, as well as science, agrees with this conception of Heraclitus.

Yet the prophet of time flow has not been able to tell us very much about the logical structure of time. Heraclitus' aphorisms supply an emotive commentary rather than a logical analysis of time. Logic has not profited from his insistence on change, whereas it did profit from Zeno's queries about change. Heraclitus' attempt to show that opposites are the same, obviously springing from his recognition that different states in time can constitute the same thing, is one of those oversimplified generalizations which are in part truistic, in part obviously false. "The way up and the way down is one and the same" is merely a formulation of the trivial fact that a relation and its converse can be used equally well to make equivalent statements; "taller" can express the same fact as "shorter", through a reversal of the order of the terms they connect. But to say that the statement "Peter is taller than Paul" means the same as the statement "Peter is shorter than Paul" would be a contradiction. In other examples of so-called identical opposites, Heraclitus merely cites instances in which extremes are at opposite ends of the same scale, like hot and cold. In others, again, he illustrates the trivial fact that a thing can have opposite relations to different other things, as in his instance of sea water, which is drinkable for fishes but undrinkable for men. An alleged logic of opposites cannot solve the problem of time and Becoming.

Heraclitus' approach to the problem of time order is naive; it is the attempt to understand time by mere reflection on meanings derived from everyday experiences. Unfortunately, this kind of approach has been regarded by many, even in our day, as the truly philosophical approach, particularly if the conclusions arrived at are formulated in an obscure and oracular language, like that of Heraclitus. But darkness of language has too often been the guise of a philosophy of trivialities mingled with falsehood and nonsense—whether it teaches the identity of opposites, the doctrine that contradiction is the root of motion and life, or the conception that the nothing is something. A clarification of the meaning of time and Becoming can be expected only if questions raised by common sense are answered with the help of scientific method. The precision of the scientific formulation, its testability by observation in combination with logic, and its far-reaching power of connecting facts from very different domains combine to form an instrument of research capable of shedding new light on problems emerging with everyday experience. The analysis of time had to be connected with the analysis of science in order to become accessible to logical clarification.

A brief consideration shows that the study of time is a problem of physics. Emotive reaction to time flow cannot determine the answer to the question: What *is* time? Subjective experience of time, though giving rise to emotional attitudes, cannot give us sufficient information about the time order that connects physical events. We know that subjective judgment about the speed of time flow is deceptive; that on some occasions, time seems to pass quickly, on others, it seems to drag, depending, for instance, on whether we are fascinated or bored. Psychologists have shown that what we call the present is not a time *point,* strictly speaking, but a short *interval* of time, the length of which characterizes the psychological threshold of time perception. An optical impression, for example, takes time to "build up"; this fact explains the perception of motion in a moving picture, which consists of static pictures shown in rapid succession. But the existence of a temporal threshold appears irrelevant to the study of time as an objective process, in the same sense as the existence of perceptual thresholds is irrelevant to the investigation of geometrical length or of sound intensity. In fact, if Zeno's criticism of the continuum is to be applicable to time, the physical process of time flow must be assumed to be independent of the psychological experience of time, which does *not* have the structure of a mathematical continuum but is "atomistic" in nature.[2] What matters is the structure of that time which controls

[2]This has been correctly emphasized by Grünbaum, *op. cit.,* p. 162.

physical events; what we wish to know is whether our emotional reaction to time is justified, whether there is a time flow, objectively speaking, which makes events slide into the past and prevents them from ever again returning to the present. In what sense does the future differ from the past? For the answer, we must turn to physics, if we wish to understand time itself, rather than mere psychological reactions to it.

It was said above that the past is distinguished from the future as the unchangeable from the unknowable. Is this distinction to mean that the future is still changeable? We would be inclined to answer in the affirmative, because of our simple daily experiences. Our control of the future, though certainly limited in extent, is often sufficient to satisfy our needs. Planned action, based on anticipation of what the future will bring us, has enabled us to turn many of its gifts to practical use. We sow and reap; we provide ourselves with shelter; we organize human society; we build machines that facilitate our daily work.

The scientist, however, might be inclined to question the belief that the future is changeable. Being unknowable, he might argue, does not imply being undetermined; perhaps the future is as determined as the past and the difference between past and future is merely a difference between knowing and not knowing. The apparent asymmetry of time would then be only a matter of knowledge and ignorance; time itself would be symmetrical, its objective nature would be the same in the direction of the past as in the direction of the future. Such conceptions suggest themselves within the scientific approach, because science has accepted the universal validity of causality.

Causal laws govern the past as well as the future. We see them at work in past facts; but we also see them confirmed by future facts which we have correctly predicted and which have later become reality. The future is not entirely unknowable. Quite a few occurrences can be predicted. Among these are the motions of the stars, the seasons, the growth of plants, animals, human beings; and certainly death is predictable. What led philosophers to question the reality of time is the fact that some undesirable future facts—in particular, death—are predictable. Why not assume that the future is as determined as the past?

Ancient philosophies, again, offer us many illustrations of this view. The idea of a causal determinism was developed in antiquity. Sometimes it appears in the form of fatalism, in which only the ends of human striving are predetermined, while the ways are left open to chance. This is an anthropomorphic form of determinism; it reminds us of a child's fear of punishment by his parents—a fear transferred to a heavenly father. But the end may be irrelevant; the determination of the future can be conceived as supplied by causal laws alone, the same

causal laws that govern the course of the stars and the growth of the
living organism. Some ancient philosophers, like Democritus, envisaged
a determinism of this kind. But a genuine causal determinism in the
modern sense of the word first arose with modern science.

The physics of Galileo and Newton revealed that many more events
can be predicted than are forseeable to common sense; and it showed
that prediction can achieve amazing quantitative precision. The use of
mathematical methods in the physical sciences has brought this kind
of success. Not only has it greatly improved astronomy and navigation,
but it has also taught us techniques which would be impossible if
causal laws did not govern physical occurrences to an incredible degree
of exactness. The steam engine and the airplane bear witness to the
determination of the future. Who would dare to step into an airplane
were he not convinced that the laws of aerodynamics formulate highly
reliable predictions?

It is no wonder that, with the progress of modern science, determinism
became an increasingly influential doctrine. Newton's physics had
unveiled the physical laws governing both celestial and terrestrial
bodies; and the same laws were supposed to reign in the realm of the
atom and control atomic motion. The mathematician Laplace did not
hesitate to assume that the precision of astronomical laws also holds
within the atomic domain. Since even human thinking and feeling merely
reflect the constellations of atoms within the brain, he concluded that
every future occurrence is as determined as the past. Only human ignorance prevents us from foretelling the future. In his famous remark about
a logically possible superman he has formulated the complete symmetry
between past and future:

> We must consider the present state of the universe as the
> effect of its former state and as the cause of the state which
> will follow it. An intelligence which for a given moment knew
> all the forces controlling nature, and in addition, the relative
> situations of all the entities of which nature is composed—if it
> were great enough to carry out the mathematical analysis of
> these data—would hold, in the same formula, the motions of the
> largest bodies of the universe and those of the lightest atom:
> nothing would be uncertain for this intelligence, and the future
> as well as the past would be present to its eyes.[3]

If this passage, which has become the classical formulation of
determinism, were true, it would spell the breakdown of a realistic

[3]Pierre S. Laplace, *Essai philosophique sur les probabilités* (first published in Paris, 1814), section "De la probabilité".

interpretation of time flow. If the future is as determined as the past, the present cannot create anything new; the causal structure which constitutes the physical world can then be held to extend either from negative infinity of time to positive infinity, or, if you like, from positive endlessness to negative endlessness. And the human mind then appears to be a mere spectator who happens to see this structure from a vantage point, the present, that slides along without being a scene of Becoming. For causal determinism, as for Parmenides, there is no Becoming.

The determinism of Newton's mechanics has received an even more explicit and precise formulation in Einstein's and Minkowski's four-dimensional space-time continuum. The three dimensions of space and the one dimension of time constitute its four axes, and physical happenings are represented as "world lines", like the lines of a diagram. The present is merely a cross section in this diagram, and it makes no difference where we put it. It is only a reference point from which we count time distances, like the year *one* from which we count our era. The structure of the space-time manifold is the same everywhere, and in both directions of time; the shape of all world lines in it is determined by mathematical laws. This timeless universe is a four-dimensional Parmenidean Being, in which nothing happens, "complete, immovable, without end . . . ; it *is* all at once, a continuous *one*". Time flow is an illusion, Becoming is an illusion; it is the way we human beings experience time, but there is nothing in nature which corresponds to this experience.

The deterministic conception of time flow may be compared to the happenings seen in a motion picture theater. While we watch a fascinating scene, its future development is already imprinted on the film; Becoming is an illusion, because it makes no difference to the happenings at what point we look at them. What we regard as Becoming is merely our acquisition of knowledge of the future, but it has no relevance to the happenings themselves. The following story, which was told to me as true, may illustrate this conception. In a moving picture version of *Romeo and Juliet*, the dramatic scene was shown in which Juliet, seemingly dead, is lying in the tomb, and Romeo, believing she is dead, raises a cup containing poison. At this moment an outcry from the audience was heard: "Don't do it!" We laugh at the person who, carried away by the emotion of subjective experience, forgets that the time flow of a movie is unreal, is merely the unwinding of a pattern imprinted on a strip of film. Are we more intelligent than this man when we believe that the time flow of our actual life is different? Is the present more than our cognizance of a predetermined pattern of events unfolding itself like an unwinding film?

Now it is true that the person in the audience is outside the happenings of the film, whereas we are part of the happenings in this world. But if our future is as determined as the "future" which a film strip holds in stock, what difference would it make if we knew it in advance? We could not change it: we would be mere spectators of our own actions, and as far as control of the future is concerned, we would not be better off than a person in the audience of a movie theater who wants to keep Romeo from committing suicide.

The paradox of determinism and planned action is a genuine one. I shall later analyze it in more precise terms; at the moment it may be sufficient to point out that if time has no direction, planned action appears incomprehensible. We plan to go to a theater tomorrow; but it seems utterly senseless to plan to go to a theater yesterday. Why do we make this distinction? We answer that we cannot change the past; we can only change the future. If we went to the theater yesterday, we did; and if we did not go, we did not; what we think of it today makes no difference. Why do we not argue in the same way about going to the theater tomorrow? In our behavior we express the conviction that time goes in one direction, because planning presupposes time flow. Can scientific analysis support this conviction?

The criticism of time flow raised by the ancients can be overcome by an improved logic of infinity. The objections resulting from a deterministic physics are much more difficult to refute. Before entering into a detailed analysis of these problems, let us study another historical attempt at a theory of time, which, in contrast to earlier approaches, was undertaken in full knowledge of the results of science and with the intention to find a way out of the dilemma of determinism or freedom. This attempt, in which Parmenides' conception of time as an illusion has been revived in modern form, is given by Kant's theory of time.

It has turned out to be the historical function of Kant's system to formulate in logical terms what earlier speculative philosophers had attempted to say in similes and mystical aphorisms. This is true not only of Kant's theory of time, but of his theory of knowledge in general. With the concept of a *synthetic a priori* Kant wanted to characterize the contribution which human reason makes to knowledge. Thus he sets forth, in a logical form, the theory that reason governs the physical world—a rationalistic theory which in Plato's theory of ideas, for instance, had been the creed of a mystic rather than a logical construction. But it is unnecessary to enter into a general treatment here;[4] we shall restrict our discussion to Kant's theory of time.

---

[4]For a general treatment of Kant's historical position, see Hans Reichenbach, *The Rise of Scientific Philosophy* (Berkeley, Calif., Univ. Calif. Press, 1951), p. 40.

The Parmenidean distinction between timeless Being and illusory time flow reappears in Kant's philosophy as the distinction between timeless things-in-themselves and time as the subjective form of things-of-appearance. It is true, Kant would have objected to putting his subjective time on a par with an interpretation of time as an illusion. However, the only difference is in emotive connotations. Logically speaking, if time is subjective, it does not describe a property of the physical world. But perhaps the term "subjective time" is more precise and more descriptive than the term "illusory time".[5]

Kant presents us with long arguments in favor of his conceptions. He says that knowledge results from an interaction between physical reality and the human mind, and that the human mind superimposes certain forms on the physical things in order to make them comprehensible and accessible to order. Such forms are space and time. In particular, they are called *forms of visualization*,[6] a term meaning that space and time determine the way we *see* things, that is, they impose conditions on our perceptions of things but have no significance for things apart from their relation to a human observer. "It is certain . . . that space and time . . . are merely subjective conditions of all our

---

[5][Great terminological difficulties arise in the attempt to discuss Kant's doctrine of the nature of time; first, because Kant was not consistent in his use of crucial terms like "subjective" and "objective", and, second, because his usages of these terms differ widely from currently accepted meanings. The author has adopted current usage, even though it sometimes leads to formulations to which Kant himself would probably have objected. The discrepancy in usage arises primarily because of Kant's metaphysics and theory of knowledge. Kant, as is well known, distinguished two realms, one of noumena or things-in-themselves, the other of phenomena or things-of-appearance. The latter realm is the only knowable one, and Kant usually restricts the applicability of "subjective" and "objective" to this realm. This is also the realm of "physical reality" in the Kantian sense. But phenomena are products partly of the human mind and are therefore subjective in a certain straightforward modern sense. Things-in-themselves, as Kant speaks of them, are unknowable, and according to the modern verifiability criterion there is no cognitive meaning in speaking of them. Yet, in Kant's conception, things-in-themselves constitute the real world which exists independently of human knowing, and such a real world would be said today to constitute objective or physical reality. This is the sense in which "objective", "subjective", and "physical reality" are used by the author when he states that, in Kant's philosophy, time and causality are subjective and that knowledge involves an interaction between the mind and physical reality. Thus, although the author uses a terminology which differs from Kant's, this difference does not affect the accuracy of his characterization of Kant's view.—M. R.]

[6]I translate Kant's term "Anschauung" by "visualization", because the usual translation as "intuition" has too many connotations which Kant did not intend.

[visualizations]", he writes in his *Critique of Pure Reason*.[7] This applies even to the happenings in my mind, revealed to me by an "inner sense". "If I myself or some other being could look at me without these conditions of sense perception . . . a knowledge would result in which the conception of time and change would not occur."[8] Time flow, change, is not a property of Being, said Parmenides; and Kant agrees, adding that time is merely the form in which we experience Being.

How does Kant come to this conception? There is no doubt that his motives are related to those of Parmenides. Like Parmenides, he wanted to escape time flow. That the fear of death determines his philosophy is obvious from his insistence on immortality. But this is not his only motive. He wanted to save freedom of will, the possibility of planned action. Parmenides was not interested in action, which presumably did not appear important to him. But Kant wanted freedom of action; perhaps not so much in order to ensure for man the possibility of improving the world, as in order to make man responsible for what he does. He wanted to establish morality; and how can human action be morally evaluated if it is not free?

Unlike the ancients, Kant faced a deterministic physics of great perfection. Newton's theory of gravitation, embodying conceptions of space and time, had greatly impressed Kant. He saw that determinism offers difficulties for human freedom, so he looked for a way out. He believed that he had discovered a way to save freedom by making time subjective—by interpreting the flow of time as the form in which we human beings experience reality, which in itself is not controlled by time. He thus teaches the *ideality* of time. In his denial of a realistic interpretation of time, Kant repeats the position of Parmenides; but his argument is derived from scientific developments which the ancients could not have imagined. In fact, if determinism, as formulated in Laplace's simile of a superman, holds, then time is subjective. There can be no time flow if the future as well as the past could be present to a superman's eyes. Kant saw correctly that determinism leaves no alternative. But Kant believed that he still could save free action; and the solution he offers us is to trade in the reality of time flow for the freedom of the will.

In order to understand the solution of the freedom problem proposed by Kant, we must include his treatment of causality in our consideration. Kant saw very well the close relationship between time and causality. He saw that we discover time order by examining causal order. The

---

[7] Immanuel Kant, *Kritik der reinen Vernunft*, author's 2d and rev. ed. (Riga, Johann F. Hartknoch, 1787), p. 66. Translation mine.

[8] *Ibid.*, p. 54.

mere sequence of perceptions, he argues, does not establish a time order of physical happenings; in fact, we may perceive events in the inverse time order of their actual happening. "Mere perception leaves the objective relation of consecutive phenomena undetermined".[9] For instance, even though the first sound impact at the site of a distant explosion preceded the light flash, we would see the light first and hear the sound later. To infer temporal order, we must know causal order.

Now for Kant causal order is as subjective as time. It is one of the categories by means of which we order our experience, but it does not express a property of things-in-themselves. Therefore physics can only tell us how the world appears to us, not how it is, independent of a human observer. And determinism refers only to the world of appearance—the world in itself is free from the rule of causal laws.

This is a strange doctrine. In order to make time flow subjective, Kant has to make even causality subjective. In order to save freedom and morality, he has to sacrifice physics as a science of objective things; it is for him merely a science of experienced things. He goes beyond Parmenides in exempting Being not only from time flow, but also from causal determination.

It is difficult to understand in what sense Kant could claim that his theory reëstablished freedom. When we act, we want to change those occurrences which take place in time; we want to change the future. But temporal happenings, Kant tells us, are subject to causality and determinism. What kind of influence, then, can we have upon them? It does not help us to assume that behind the experienced things there are other things not controlled by physical laws. We want to change those things that *are* subject to causality and *are* perceived in our experience; we even use causal laws to control them, knowing very well that if things did not conform to such laws we would be hopelessly lost in our attempts to plan the future. What is a philosophy good for, if it evades answers to questions about what men can do, by telling us that there is another realm of Being which we certainly cannot control? Kant's philosophy of subjective time and subjective causality is a form of escapism. It does not solve the paradox of freedom and determinism; it does not clarify the experience of time flow; it cannot account for the distinction between past and future, between the unchangeable and the realm of what we hope can be changed.

The distinction between the past and the future, the objective interpretation of time as a process of Becoming and not merely a form of human experience, has strong support in common sense. It would not

---

[9] *Ibid.*, p. 234.

be easy to acquiesce in a philosophy which regards such conceptions as illusions. Yet the convictions of common sense cannot be accepted without criticism by the philosopher. The problem of time cannot be solved by an appeal to intuitive knowledge, which tells us that there must be a process of Becoming, making planned action possible. Reliance on so-called intuition has too often turned out to be misleading. There is such a thing as an escape into common sense as well as an escape into metaphysical speculation. Neither of them offers answers acceptable to those who look for an unprejudiced approach, guided by logical analysis in combination with observation.

This remark applies to such attempts as were made by Bergson, who claims to have constructed a philosophy for which Becoming is the essence of time. Like his predecessor in antiquity, Heraclitus, Bergson is one of those who react to time flow with a positive emotional attitude, who see in change the very element of life, and whose optimistic approach to the problem of time flow is not troubled by the threat of death. To Bergson, as to Heraclitus, the idea of Becoming offers emotional reward.

Such an attitude may be a good help in starting a logical investigation, but it cannot replace it. Bergson's spirited appeal to intitution, to the "immediate data of consciousness", cannot establish a theory of time. He argues that the physicist has misunderstood time—that he has "spatialized" time by treating it like a dimension of space. Real time, or "duration", he says, can only be understood by immediate awareness, which reveals to us that time is an act of Becoming. Such arguments do not say very much to someone who wants to know whether he can trust his intuitions. If a man is under posthypnotic suggestion, his intuition tells him that he is free; but we know he is not. If intuition tells us that the future is generated by "creative evolution", I should like to know, not only whether this is true, but also what this phrase means. But the answer cannot be found by another appeal to intuition. The meaning of Becoming can be clarified only by logical analysis; and this analysis is to be based on all we know about the physical world, including ourselves. An act of vision is no substitute.

There is no other way to solve the problem of time than the way through physics. More than any other science, physics has been concerned with the nature of time. If time is objective the physicist must have discovered that fact, if there is Becoming the physicist must know it; but if time is merely subjective and Being is timeless, the physicist must have been able to ignore time in his construction of reality and describe the world without the help of time. Parmenides' claim that time is an illusion, Kant's claim that time is subjective, and

Bergson's and Heraclitus' claim that flux is everything, are all insufficiently grounded theories. They do not take into account what physics has to say about time. It is a hopeless enterprise to search for the nature of time without studying physics. If there is a solution to the philosophical problem of time, it is written down in the equations of mathematical physics.

Perhaps it would be more accurate to say that the solution is to be read between the lines of the physicist's writings. Physical equations formulate specific laws, general as they may be; but philosophical analysis is concerned with statements *about* the equations rather than with the content of the equations themselves. It was mentioned that the question of determinism has a great bearing upon the problem of time, but there is no physical equation that formulates determinism. Whether determinism holds is a question *about* physical equations; it is the question whether certain equations supply strict predictions and cover all possible phenomena. For such reasons, the philosophical investigation of physics is not given in the language of physics itself, but in the metalanguage, which speaks *about* the language of physics.

It is well known that the determinism of classical physics has been abandoned in modern quantum physics, and it will be most important to investigate the implications which the turn toward indeterminism entails for the problem of time. But there are other investigations which must precede the study of quantum mechanics. In thermodynamics, physics has been explicitly concerned with the problem of time flow; it is a gross misunderstanding of physics to say that it has "spatialized" time. The specific nature of time, as different from space, has found an expression in certain very fundamental physical equations. It will be seen that the contribution of quantum mechanics to the time problem can be understood only after a study of the thermodynamical approach. And it will turn out that physics can account for time flow and for Becoming, that common sense is right, and that we can change the future. But the proof requires the use of scientific method. Even the meaning of the terms "time" and "becoming" can only be understood by a common sense which has assimilated the results of scientific thought.

CHAPTER II | THE TIME ORDER OF MECHANICS

● *2. The Qualitative Properties of Time*

If we wish to establish the thesis that time can be accounted for in terms of physics, we must first list those properties of time which make up its peculiar structure in the conception of the physical world that governs everyday life.

Let us begin by distinguishing *quantitative* from *qualitative* properties of time. In measuring time by the help of clocks we make use of its quantitative, or *metrical*, properties. Such measurements concern the determination of time distances of equal length, represented, for instance, by two consecutive hours; and, in addition, the determination of simultaneity, that is, of equal time values for spatially distant points. The theory of the metrical properties of time has been developed in great detail in modern physics—in particular, in Einstein's theory of relativity and in its logical analysis—and we therefore need not study it in the present investigation, but can, instead, refer the reader to other publications.[1]

Our inquiry shall be focused on the qualitative, or *topological,* properties of time. They are more fundamental, in that they hold independently of specific procedures of measurement and remain unchanged even if the forms of measuring time are varied. They comprise those properties which confer upon time its specific nature as different from space and which account for our emotive attitude toward time, as was explained in the preceding section. We shall try to formulate these qualitative properties in several statements. Since we are at the beginning of our examination, we shall not be too insistent upon precision. We want to collect material to work on, and we leave the elaboration of more precise formulations to later sections of our study.

The most obvious properties of time can be formulated as follows:

Statement 1. Time goes from the past to the future.

This statement refers to the flow of time; it expresses what we call *becoming.* Time is not static; it moves. When we speak of the flow of time, we usually regard it as the flux of some objective entity which we perceive and the continual escaping of which we cannot prevent. Yet there remains the possibility that somehow the flow of time is connected with the structure of human consciousness, and that, though derived from an objective root, it is the common product of an objective factor and a subjective one. We shall be able to answer this question only after examining more precisely the physical root of time.

Statement 2. The present, which divides the past from the future, is now.

This statement appears rather puzzling. In one sense, it seems to be a trivial tautology, because the terms "present" and "now" mean the same thing. In another sense, it is strange and enigmatic; for what is the meaning of "now"? Does the word "now" refer to a time point physically distinguished from other time points? Or does it merely express our subjective approach to time—the vantage point from which we see time, so to speak, comparable to the space point here, from which we see the things around us in a certain spatial perspective? But we can choose the center of spatial perspective; for instance, we can walk around the table and look at it from the other side. Can we select the Now? There are restrictions on that. We can decide to wait for tomorrow in order to have a different Now; but we cannot make

---

[1]See Hans Reichenbach, "Bericht über eine Axiomatik der Einsteinschen Raum-Zeit-Lehre", *Phy. Zeitschr.,* Bd. 22 (1921), pp. 683-686; *Axiomatization of the Theory of Relativity* (Berkeley and Los Angeles, University of California Press, 1969); and *Philosophie der Raum-Zeit-Lehre* (Berlin and Leipzig, W. de Gruyter & Co., 1928; English edition published in 1957 by Dover Publications, Inc., New York).

yesterday our Now. And even in postponing the selected Now for tomorrow we cannot escape the fact that we do so *right now,* that is to say, that another Now imposes itself upon us as the one in which the act of decision takes place, the act in which one decides to postpone the intended Now.

These considerations lead to the following statements concerning the past:

Statement 3. The past never comes back.

This statement appears to be closely connected with the flow of time, that is, with statement 1. It expresses the fact that the flow goes in the direction of a line that nowhere intersects itself and thus can be conceived as a straight line, progressing from negative infinity to positive infinity. The conception of time as a one-dimensional or linear continuum has its root in statement 3.

Although the meaning of the terms "past" and "future" appears quite obvious to us, it will be useful to look for specific differences between the past and the future. The following three statements are intended to formulate such differences.

Statement 4. We cannot change the past, but we can change the future.

The first part appears obvious, but the second requires some qualification. Our ability to change the future is very limited. We can prevent our car from colliding with another by stepping on the brake; and we can change barren lots into residential quarters. But we cannot control cosmic events, or the weather, or earthquakes; and we are rather poor at controlling human society, which continues to drift from crisis into crisis and from war into war. The statement means, more precisely speaking: There are some future happenings which we can control, but there are no events of the past which we can change.

We might be tempted to formulate another difference by saying that the future is unknown, whereas the past is known. Such a statement, however, would be obviously false. Some events of the future are well known, such as astronomical events, or the fact that there will be general elections in the fall. And many events of the past are unknown; if they were known, historians and geologists would have an easier task. The difference between the past and the future is not a difference between knowing and not knowing; it is to be formulated with respect to the way in which we acquire such knowledge.

Our knowledge of the past is based on *records*, whether they be documents written by a chronicler, or fossils included in geological strata, or traces of blood on a garment. In contrast, it appears absurd to speak of records of the future. Records are small by-products of comprehensive events, the total effects of which include many other and more important happenings. The chronicle is a by-product of a war that killed thousands of people; the fossils are by-products of extensive geological events; the traces of blood on the garment may be by-products of an act of murder. Such isolated data are not enough, if we wish to predict the future.

The happenings of the future can be foretold only on the basis of comprehensive information covering the total occurrence. If we wish to predict whether there will be a war, we have to know the state of armament of various countries, their political aspirations, and many other things. If we wish to predict whether a certain area of the ground will later be covered by water, we have to make measurements concerning changes of level over much wider areas. In order to predict an act of murder, we would have to have a detailed knowledge of a certain person's state of mind.

For this reason, we have registering instruments only for past events, and not for future ones. We saw that such instruments are controlled by *partial* effects of comprehensive occurrences, yet effects of such a kind that they allow us to infer that the larger occurrence took place. It appears nonsensical to speak of a registering instrument of future events. Consider, for instance, a registering barometer. It contains a needle that is raised or lowered by the pressure of the surrounding air, which compresses the walls of an evacuated vessel; and the position of the needle is indicated on a slowly rotating drum by the flow of ink. This insignificant effect was perhaps the by-product of a storm that felled trees and took roofs off houses. Yet from the mark on the drum we can infer only that there was this amount of air pressure at this place. Conversely, when we wish to predict the air pressure at the same place tomorrow, it is not sufficient to make measurements merely at the place considered. We would have to use meteorological data taken over a wide geographical area in order to predict the atmospheric pressure in one location, data including much more information than barometric readings alone.

Only in exceptional situations may it be possible to predict the future from isolated indications. Such a situation occurs when certain observations make possible an inference toward the past revealing that a comprehensive process is going on, the effects of which are predictable. For instance, cracks in the wall may enable us to predict that a house will break down. A governmental order to evacuate all women and children from a certain area may indicate that a war is imminent. In such cases, the total cause that produces the predicted effect can be inferred from a few isolated observations. These data are first used as records for making an inference toward the past, toward a comprehensive cause; and then a prediction is made by the use of the knowledge of the total cause. And even if such a prediction is possible, it can be made only in general terms and will never supply the detailed information attainable concerning recorded events. We may be able to predict that the house will break down, but we cannot foretell the exact places where the broken parts will be found, information which can later be easily stated in records. We may be able to predict a war, but we cannot predict which persons will be killed, information that later records will give us. Even those who insist that such detailed predictions are possible in principle, though not in practice, will admit that these forecasts cannot be based on a few isolated indications. This difference concerning details springs from the peculiar difference between inferences regarding the future and inferences concerning the past. *Predictions* require a knowledge of the total cause; *postdictions*, or statements about past events, can be based on partial effects, on records.

We formulate this result as follows:

Statement 5. We can have records of the past, but not of the future.

Finally, we formulate now a difference between past and future which appears to sum up the differences expressed in statements 4 and 5:

Statement 6. The past is determined; the future is undetermined.

If we were asked what we mean by the terms "determined" and "undetermined", we should presumably feel somewhat uneasy. We might argue that the past consists of established facts, whereas the future does not; and we might explain that an established fact is something which we cannot change, and which can be recorded, thus referring back to statements 4 and 5. Indeed, it is questionable whether the meaning of statement 6 goes beyond that of statements 4 and 5.

This is one of the questions we shall have to investigate. But there are further questions to be raised. We have been making a preliminary survey of the qualitative properties of time; and many of the terms used require an examination of their meanings. Is it clear what we mean when we state the sheer fact that we can change certain future events? Is the word "record" sufficiently defined? And what do we mean by the flow of time, referred to in statement 1?

Our analysis of time, therefore, is concerned with a clarification of meanings. It is a problem of *explication;* that is, we have to construct precise concepts to replace the vague ones thus far used. In an explication, the vague concept is called the *explicandum,* whereas the precise concept is called the *explicans.* The aim is to find an explicans which, put into the place of the explicandum, justifies the use of the statement in the context of human behavior.

Explication problems play a great part in the investigations of scientific philosophy. It may be recalled that such questions as "What is probability?" "What is a causal law?" "What is logical necessity?" and "What is the mind?" are problems of explication; and the philosophical discussion of such problems should be approached from this point of view. An explication can never be proved to be strictly correct, for the very reason that the explicandum is vague and we can never tell whether the explicans matches all its features. We can merely require that an explication be adequate, that is, that the explicans correspond, at least qualitatively, to the usage of the term in conversational language, and that if the explicans is put into the place of the explicandum, most sentences of conversational language do not change their truth values. If this requirement is satisfied, we can regard the explication as a proposal to use the new term instead of the old one. This replacement will help us not only to arrive at precise meanings, but also to formulate relations concerning the concept which otherwise would remain unknown. Explication is therefore the method by the use of which we eventually succeed in understanding the meaning of a concept too involved to be accessible to direct understanding.

The explication of the concept of time, with its derivatives like time direction, time flow, and so on, is of particular significance because of the major part which time plays both in everyday experience and in the sciences. It will be seen that the explication of time requires a study of some fundamental chapters of physics, and that it will clarify the foundations of physics as well as the foundations of philosophy. That these investigations are of vital importance to philosophy is obvious. For, so long as the nature of time is not understood, the philosopher cannot say that he is able to give an account of physical reality or of human knowledge about it.

• 3. The Causal Theory of Time

How can we find a suitable explication of time? It is clear that it can be sought only by a study of the relationships of causality. The close connection between time order and causal processes will be evident from the preceding discussion. It will be the aim of the present investigation to show that this is more than a mere coincidence, that time order is *reducible* to causal order. Causal connection is a relation between physical events and can be formulated in objective terms. If we *define* time order in terms of causal connection, we have shown which specific features of physical reality are reflected in the structure of time, and we have given an explication of the vague concept of time order. We shall show, in particular, that statements 1-6 in §2 can thus be given a precise meaning, a meaning in which they are true, although the adoption of this meaning requires us to abandon certain familiar connotations and commits us to a probability interpretation of causality.

The idea of reducing time order to causal order was first conceived by Leibniz.[1] His brief outline of such a reduction bears the imprint of the genius who anticipated the conception of the relativity of space and time. However, the causal theory of time could not be definitively established before Einstein had completed his theory of relativity. The decisive argument in favor of defining time order in terms of causal order derives from Einstein's criticism of simultaneity. It is well known that the Lorentz transformations, which express Einstein's special principle of relativity, permit the reversal of the time order of certain events, namely, of those which cannot be connected by causal chains. Time order is invariant under the Lorentz transformations only if the events in question can be connected by signals, that is, by causal chains. It follows that if time order were more than causal order the Lorentz transformations and Einstein's relativity could not be accepted. This point cannot be emphasized too much: those who reject the causal theory of time, who insist that time order has a meaning independent of causal order, are compelled to deny physical significance to the Lorentz transformations; and the theory of relativity becomes for them a play with symbols, most of which do not refer to any physical reality.[2]

In view of such considerations, I have developed in earlier publications[3] a causal theory of time. It is not planned to summarize my results here. In these earlier investigations, the relation of cause and effect was assumed as a primitive term, and it was shown that temporal order, and even spatial order, can be reduced to this primitive relation. In the present book, I wish to study the cause-effect relation in itself; that is, to treat it no longer as primitive, but to reduce it to other relations. Which other relations will be required will be seen in the course of the presentation. This analysis will enable us to solve the problem of the direction of time, a problem that cannot be solved in the frame-

---

[1]See the presentation in Hans Reichenbach, "Die Bewegungslehre bei Newton, Leibniz und Huyghens", *Kant-Studien*, Bd. 29 (1924), p. 421.

[2]See also the remarks on p. 39 of the present study.

[3]See Hans Reichenbach, "Bericht über eine Axiomatik der Einsteinschen Raum-Zeit-Lehre", *Phys. Zeitschr.*, Bd. 22 (1921), pp, 683-686; *Axiomatization of the Theory of Relativity* (Berkeley and Los Angeles, University of California Press, 1969) ; and *Philosophie der Raum-Zeit-Lehre* (Berlin and Leipzig, W. de Gruyter & Co., 1928).

work of Einstein's theory of relativity, because it requires a transition from strictly causal relations to probability relations.

A remark of a logical nature must be added. The discussion of the problem of time has greatly suffered from the confusion of two concepts, from neglecting the distinction between *order* and *direction*.

The points on a straight line, which is infinite on both sides, are arranged in a certain order; but the line does not possess a direction. When the points are in a linear order, or serial order, they are governed by an asymmetrical and transitive relation. Referring to figure 1, we may say, for instance, that A is to the left of B; then B is not to the left of A, but to the right of A. This fact expresses the asymmetry of the relation *to the left of.* The relation *to the right of* is called the converse of *to the left of.* Furthermore, if A is to the left of B, and B to the left of C, then A is to the left of C; this fact expresses the transitivity of the relation. It is well known from the theory of relations that every asymmetrical, connected, and transitive relation establishes a serial order. When we say that a line, though serially ordered, does not have a direction, we mean that there is no way of distinguishing structurally between left and right, between the relation and its converse. In order to say which direction we wish to call "left", we have to point to the diagram; or we may give names to points and indicate the selected direction by the use of names. Had we decided to call "right" what we called "left", and vice versa, we would not notice any structural difference; that is, the relation *to the left of* has the same structural properties as the relation *to the right of.*

It is different in the case of the linear continuum of negative and positive real numbers, which we can map on a straight line. The numbers are governed by the relation *smaller than*, which is asymmetrical, connected, and transitive, like the relation *to the left of;* therefore the numbers have an order. But in addition, the relation *smaller than* has a direction; that is, it is structurally different from its converse, the relation *larger than.* This is seen from the fact that we can distinguish between negative and positive numbers in the following way. The square of a positive number is positive, and the square of a negative number is also positive. We therefore can make this statement for the class of real numbers: Any number which is the square of another number is larger than any number which is not the square of another number. We thus have defined the relation *larger than*, and with it, the relation *smaller than*, in a structural way, without having to point to a diagram.

Applying these results to the problem of time, we find that time is usually conceived as having not only an order, but a direction. The relation *earlier than* is regarded as being of the same kind as the relation *smaller than*, and as not being undirected like the relation *to the*

Fig. 1. The order of points on a line.

left of. This means that we believe that the relation *earlier than* differs structurally from its converse, the relation *later than.* To formulate this structural difference is a major problem of our inquiry; the method used for numbers, of course, does not help us for time points. We introduce numbers for time points, rather, after having defined a direction for time. The distinction between past and future is intended to express the direction of time; and the six statements given in the previous section seem to formulate a direction, not merely an order, of time. Whether this is true for all of them remains to be seen.

• *4. Causality in Classical Physics*

Since we wish to define time in terms of causality, we must now investigate whether the cause-effect relation possesses a direction, or at least, an order.

In the experiences of everyday life we take it for granted that the cause-effect relation is directed. We are convinced that a later event cannot be the cause of an earlier event. But when we are asked how to distinguish the cause from the effect, we usually say that of two causally connected events, the cause is the one that precedes the other in time. That is, we define causal direction in terms of time direction. Such a procedure is not permissible if we wish to reduce time to causality; and we therefore must look for ways of characterizing the cause-effect relation without reference to time direction. Let us see whether the laws of physics supply a criterion of this kind.

The view that such a criterion exists has been criticized by means of the following consideration. The laws of physics state functional relationships. They assert that, if certain physical quantities $x_1 \ldots x_n$ have certain values, another quantity x_{n+1} has a determined value. Thus they can be written in the form[1]

$$x_{n+1} = f(x_1 \ldots x_n) \ .$$
(1)

---

[1]Formulas are numbered by section. Reference within the same section is made by formula number only, but when reference is made to a formula in another section, the number of the section is prefixed to that of the formula. To illustrate, the reference (5, 2) means formula (2) of §5, but (2) alone means formula (2) of the current section.

In general, such equations can be solved for any of the variables contained in it; for instance, we can write

$$x_1 = g\left(x_2 \ldots x_{n+1}\right) \ , \tag{2}$$

g being a function resulting from f according to rules for mathematical equations.

It has been argued that, because of this convertibility of functions, physical laws do not define a direction or an order. Functional relationship, it is said, is a symmetrical relation; if y is a function of x, then x is a function of y. Now there is no doubt that the latter statement is true, that functional relationship is, indeed, a symmetrical relation. But the question remains whether we have to conclude that causality is a symmetrical relation. The relation of causality may be of such a nature that it is not exhaustively characterized by the general concept of functional relationship.

Conceptions of this kind have been in ill repute because they have been advanced within metaphysical philosophies. Philosophers who maintain that observation cannot inform us about the ultimate laws of the physical world regard causality as a relationship of an a priori nature, that is, as a relation the form of which is found through reasoning alone. They would argue that functional relationships constitute merely that part of knowledge which observation can teach us, but that reason adds to this knowledge an a priori chapter in which causality is seen to establish a sort of mysterious thread that connects the effect to the cause and makes it the product of the cause. For such philosophies, causality is a directed relation.

I will leave to other presentations the criticism of such views, which are no longer taken seriously in scientific philosophy. However, I should like to make it clear that the conception of a directed causality does not necessarily entail any metaphysics; that a distinction between causality and mere functional relationship can very well be carried through within an empiricist philosophy, which insists that all statements of a synthetic nature be verifiable in terms of observables. More precisely speaking, causality may be a very specific functional relationship, which allows for the definition of a direction. We used above the example of the relation *smaller than* between real numbers; in this instance, too, we define a directed relation in terms of functional relationships without resorting to metaphysical conceptions.

It is true that there are certain functional relationships of physics which are symmetrical, but they can easily be shown to be compatible with the assumption that causality is a directed relation. It is a familiar logical theorem that from a serial relation we can always construct a symmetrical relation. Similarly, we define a relation of causal connection as follows:

DEFINITION. An event A is *causally connected* with an event B if A is a cause of B, or B is a cause of A, or there exists an event C which is a cause of A and of B.

It is this relation of causal connection, which is obviously symmetrical in A and B, that we use in the symmetrical functional relationships of physics. Consider, for instance, the law of Boyle and Mariotte for perfect gases:

$$p \cdot v = R \cdot T \cdot \frac{M}{m} . \tag{3}$$

It says that the quantities of pressure (p), volume (v), molecular weight (m), mass (M), and temperature (T) are in a certain functional relationship, the formulation of which requires a certain constant, R, which is independent of the substance. This law holds for all kinds of changes in the values of the quantities, but it does not tell us which change is the cause of which other change, or which quantity changes first so as to make the others vary. When we experiment with gases, we know very well what is the cause of the observed changes, but we know it from other considerations than the relation laid down in law (3). When we inflate an empty inner tube of a tire, the rise in pressure and mass of the air is the cause of a rise in the volume; when the piston in the cylinder of the car's engine moves upward, the decrease in volume is the cause of the increase of the pressure in the enclosed gasoline vapor, while the mass remains constant. During these processes, the numerical values of the quantities are controlled by (3); however, the relation (3) does not tell us what process is going on. This equation formulates merely a relation of causal connection, leaving open the question of the cause of the change.

Many physical laws are of this type and refer merely to causal connection, without defining a direction or an order. The law of the conservation of energy and Ohm's law for electric circuits are of this symmetrical kind. When we wish to find a directed cause-effect relation, we have to look for other kinds of laws.

This consideration leads us to a new approach. We have to look for laws that describe physical processes, and do not merely lay down causal connections. Such laws exist. They can be grouped into two different kinds: the first kind consists of laws describing mechanical processes; the second, of laws describing thermodynamical processes. The difference between these two kinds of processes is well known: mechanical processes are reversible, whereas thermodynamical processes, apart from certain exceptions, are irreversible. Let us attempt to define this distinction in precise terms.

A process Q going on in time represents a series of states ordered in the direction of time. When we describe the process in this form, we speak of a description in positive time. We can construct a second description by using negative time; we then number the time points in reverse, or in other words, introduce a new time t' connected with the original time t by the relation $t' = -t$. In the description of the process Q we now replace t by $-t'$; the resulting description will be called the converse description, or description in negative time. It describes the same process Q, but in a different language. Obviously, we can construct a converse description for every process.

We can now use the converse description and give it a different meaning: we regard it as a description in positive time. This is easily done simply by canceling the prime mark. The question arises whether there is a process Q^* which conforms to this description; or, more precisely speaking, whether the new description denotes a process Q^* which is compatible with physical laws. If it is, we may say that the process Q is reversible, and call Q^* its reverse process. If it is not, we say that Q is irreversible, or that the reverse process does not occur. The reverse process is, so to speak, the mirror image in time of the original process.

The use of an example which admits of a simple mathematical treatment will illustrate these considerations. Let the process Q consist in a ball being thrown from A to C (fig. 2) in the direction of the solid arrow. Its motion is given by the equations

$$x_1 = a_1 t - \frac{g}{2} t^2, \qquad x_2 = a_2 t, \tag{4}$$

where a_1 and a_2 are the components of the velocity with which the ball is thrown from A. Introducing negative time t' by the relation $t = -t'$, we find

$$x_1 = -a_1 t' - \frac{g}{2} t'^2, \qquad x_2 = -a_2 t'. \tag{5}$$

This is the description of Q in negative time. We now construct the description of the reverse process Q^* of Q by omitting the prime mark in (5):

$$x_1 = -a_1 t - \frac{g}{2} t^2, \qquad x_2 = -a_2 t. \tag{6}$$

Here t is, as in (4), positive time. Obviously, the process described in (6) is compatible with mechanical laws and does occur; it is exemplified by a ball which is thrown from C to A, in the direction of the broken arrow. Throwing a ball is a reversible process.

The reason that mechanical processes are reversible has often been pointed out: the differential equations of mechanics are of the second

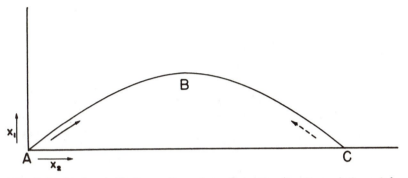

Fig. 2. Path of a ball thrown from A to C in the direction of the solid arrow. The broken arrow indicates the motion in reversed time.

order without first time derivatives. If $f(t)$ is a solution of such an equation, then $f(-t)$ is likewise a solution, because in the double differentiation a double multiplication by $\frac{d(-t)}{dt} = -1$ occurs, which leaves the result unchanged. For instance, both (4) and (6) satisfy the differential equation

$$m \cdot \frac{d^2 x_1}{dt^2} = -m \cdot g \ , \tag{7}$$

which formulates Newton's famous principle that the force (right-hand side) equals mass times acceleration (left-hand side). The minus sign in the expression representing the force results because we have given positive values to the x_1-coördinate in the direction against gravity.

We give this explanation in such detail because we need very precise concepts when we treat the problem of time. A converse description is possible for every process; but it is not true that for every process there is a reverse process. Irreversible processes are processes such as the mixing of gases or liquids and the passage of heat from bodies of higher to bodies of lower temperature. Processes of this nature are characterized by an increase of entropy, that is, by a tendency toward the disappearance of existing differences; and the reverse process is excluded, because it would represent a decrease of entropy. Processes of this kind will be studied in §7.

The use of converse descriptions, that is, descriptions in negative time, is a convenient device for the study of time problems. Since it is always possible to construct a converse description, positive and negative time supply equivalent descriptions,[2] and it would be meaningless

[2]For a discussion of the theory of equivalent descriptions, see Hans Reichenbach, *The Rise of Scientific Philosophy* (Berkeley, Calif., Univ. Calif. Press, 1951), pp. 133, 136, 147, 180, 266.

to ask which of the two descriptions is true. However, it is a question of truth whether a certain reverse process exists. Furthermore, it is a true statement that positive time is the time of our subjective experience; we shall analyze this statement in a later chapter.

Negative time is very well illustrated by the use of motion picture films reeled in reverse. Processes which in this way of reeling appear compatible with physical laws are reversible; those which appear incompatible are irreversible. Mechanical processes like the throwing of balls and the lifting of a vase from a table appear quite natural in the reversed movie. In contrast, the burning of cigarettes, the pouring of cream into coffee, the breaking of pottery, and so on, lead to incredible happenings in reversed time. Such occurrences represent irreversible processes.

In the usual discussion of problems of time it has become customary to argue that only irreversible processes supply an asymmetrical relation of causality, while reversible processes allegedly lead to a symmetrical causal relation. This conception is incorrect. Irreversible processes alone can define a direction of time; but reversible processes define at least an order of time, and thereby supply an asymmetrical relation of causality. The reader is referred to the discussion of the relation *to the left of* (see §3). The correct formulation is that only irreversible processes define a *unidirectional* causality.

Our statement that the laws of mechanics, though leading to reversible processes, supply an asymmetrical causality and thus an order of time, requires a more detailed analysis, which will be given in the next section.

• 5. *The Causal Definition of Time Order*

When we wish to find out what order properties of time can be defined by means of reversible processes, we must look for properties that remain invariant for a reversal of time direction. Such properties exist, indeed. They are represented by those order relations that are expressed by the word "between". The between-relation is symmetrical in its outside points: if B is between A and C, it is also between C and A. For this reason, reversible processes can define a linear order.

We saw that the motion of a ball from A to C, represented in figure 2, does not define a time direction, because the reverse motion from C to A is likewise compatible with the laws of mechanics. For both processes, however, the passing of the ball at B is temporally between the passing at A and the passing at C. Therefore, the laws of mechanics can very well inform us about temporal between-relations. They tell us that there is a causal line from *A* by way of *B* to *C*, or from *C* by way of *B* to *A*, and thus leave merely the direction of the line undetermined.

The significance of this result will be even more obvious when we use two processes. Figure 3 represents two balls being thrown simultaneously from *A*, one to the right as before, and another one to the left. This is the description in positive time, indicated by the solid arrows. In negative time, the two balls would travel in opposite directions, as indicated by the broken arrows. Both interpretations are

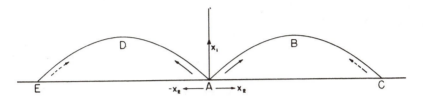

Fig. 3. Paths of two balls thrown from *A*, one to *C* and the other to *E*, in the respective directions of the solid arrows. The broken arrows indicate the motions in reversed time.

compatible with the laws of mechanics. Thus we know that we have here two causal lines, which either both start at *A*, or both end at *A*. However, we cannot interpret one ball as moving in the direction of the broken arrow and the other ball as moving in the direction of the solid arrow. Therefore, though we do not know the direction of the lines, we do know a relation between the directions. We know that the two lines cannot be combined into one; they are *counterdirected*. The latter term expresses an order property, which holds for both directional interpretations.

This consideration requires some comment. A simultaneous motion of one ball in the direction of the broken arrow and of the other in the direction of the solid arrow is, of course, compatible with physical laws; however, it is not compatible with those observations which led to the two preceding interpretations. For a clarification of this question, let us assume that the balls collide at *A*; the motion at *A* is then like that described by the diagram of figure 4, which represents a close-up, so to speak. We will assume that strict coincidences, as well as approximate coincidences, are observable. This is a continuity assumption; it means that the approximate coincidences in the space-time neighborhood of a strict coincidence are defined observationally just as well as strict coincidence. Referring to figure 4, we then can say: the practical coincidences 1-4, 2-5, and 3-6 are observed, and there remains

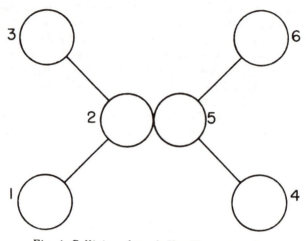

Fig. 4. Collision of two balls, illustrating the
local comparability of time order.

only the problem of their time order. If one ball in figure 3 moved in
the direction of the broken arrow and the other in that of the solid one,
we would observe the coincidences 1-6, 2-5, and 3-4, the time order of
which would still be left open. We now regard it as a matter of observation that the set 1-4, 2-5, 3-6 occurred, and not the other one. For instance, we may assume that, in the neighborhood of a strict coincidence,
it is possible to observe the direction of the line which connects
the centers of the balls. An observer may look at the balls from the
side; if the ball in position 1 covers the other ball for his eyes, the
other ball is in position 4, and not in 6. Such observations are as possible as observations of coincidences; in fact, ascertaining the coördinates x_1 and x_2 of a ball in a certain position is done similarly, by
sighting the ball in front of a scale. Observations of the kind explained
compel us to interpret the occurrences either in the sense of the solid
arrows or in the sense of the broken arrows, but they exclude a mixed
interpretation.

It should be noted that this distinction is not a matter of the laws of
mechanical collision. If one ball moved in the direction 1-2-3 and the
other in the direction 6-5-4, the laws of collision would lead to the
same diagram as that given in figure 4. But the motion process would
lead to different observables. The extension from strict coincidences
to practical (including approximate) coincidences is indispensable if
we are to satisfy continuity postulates; otherwise, the same volume
element could be given two opposite time directions.

In other words, if two processes occur in spatiotemporal juxtaposition we regard a comparison of their time directions as possible. The statement that, in a given description, both processes have the same time direction, is regarded as observationally verifiable. We shall speak here of the *local comparability of time order.* Without continuity assumptions of this kind the concept of coincidence could not be defined, because we always observe practical coincidences only and introduce strict coincidence by a procedure of schematization.

Assumptions of this kind are made in the usual treatment of mechanical problems. The differential equations of mechanics alone do not determine the equation of the motion; in addition, we have to use boundary conditions. Using equations (4, 4),[1] we would say, assuming that the two balls are thrown with equal velocity, that their motion is given by the equations:

Ball ABC: $\qquad x_1 = a_1 t - \dfrac{g}{2}t^2$, $\qquad\qquad x_2 = a_2 t$;

$$\tag{1}$$

Ball ADE: $\qquad x_1 = a_1 t - \dfrac{g}{2}t^2$, $\qquad\qquad x_2 = -a_2 t$.

The reverse process results when we substitute "$-t$" for "t" in *both* equations:

Ball CBA: $\qquad x_1 = -a_1 t - \dfrac{g}{2}t^2$, $\qquad\qquad x_2 = -a_2 t$;

$$\tag{2}$$

Ball EDA: $\qquad x_1 = -a_1 t - \dfrac{g}{2}t^2$, $\qquad\qquad x_2 = a_2 t$.

We regard both solutions as compatible with the observed boundary conditions, unless an irreversible process were to single out one of the solutions. But putting "$-t$" for "t" in the equation of only one ball would lead to a solution which is incompatible with the observables.

This consideration can be summarized as follows. Neither the laws of mechanics nor mechanical observables give us a direction of time, unless such a direction has been defined previously by reference to some irreversible process. For instance, if the velocity of a body is regarded as an observable, its direction must be ascertained by comparison with some temporally directed process, such as the time of psychological experience, which is derived from the irreversible processes of the human organism. But if no such standard is used, we cannot regard a velocity as an observable. We can merely derive its

---

[1]Interpret (4, 4) as §4, formula (4). See the first footnote in §4 for an explanation of the system of citing formulas.

value from other observables, which, however, leave the sign of the velocity undetermined.

This is the case if a velocity is computed from the observation of spatial positions. For instance, the path and the initial velocity of a bullet can be computed if at least three points of the path are given, say, by holes in walls. Or a velocity can be inferred from the observation of a centrifugal force. But the sign of the velocity remains unknown as long as no irreversible processes are adduced for the interpretation of the observations. This is what is meant when we say that neither the laws nor the observables of mechanics permit us to distinguish between the reverse process and the original one. But the relative order of two mechanical processes can be ascertained; this is the meaning of the principle of the local comparability of time order.

The determination of counterdirected lines leads to negative statements concerning the time order of events belonging to different lines. Thus, we conclude that the passage at A is not temporally between the passage at D and the passage at B. Either the passage at A is temporally before the two other passages or it is after them. We see that mechanical laws can establish and can deny the existence of temporal between-relations. Applying similar methods to further events, we thus can construct a *causal net* which, as a whole, has a *lineal order*. This term means that, if a direction is assigned to one line, a direction is determined for each line. If we reverse the direction for the first line, all other line directions are reversed. In figure 5, the directions are indicated for one time interpretation; for reverse time all arrows would be reversed. It is an important fact that a general lineal order can be constructed by the use of the reversible processes of mechanics, if only a local comparability of time order is assumed.

Now we come to formulate a very important property of the causal net, which must be regarded as the result of many experiences. Let us assume that we have selected a direction for one line; as we know, a direction is then assigned to all lines. We can now travel along the lines, starting at one point and always following the direction of the arrows. If several arrows depart from one point, we select one as we like. A combination of lines traveled through in this way may be called a *causal chain*. Traveling along causal chains we now make the discovery that we never return to the starting point; or, to put it another way, that *there are no closed causal chains*.

The nonexistence of closed causal chains is a general property of the net; we shall say that the net is *open*. Obviously, if this property holds for one choice of the line directions, it holds likewise when all directions are reversed. This means that the openness of the net is an order property, not a direction property.

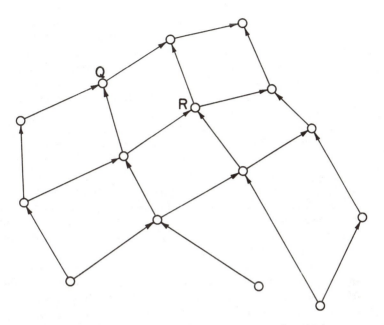

Fig. 5. A causal net. *Q* and *R* indicate events for which
no time order is determined.

It should be kept in mind that the openness of the causal chains
represents an empirical fact and cannot be regarded as a logical necessity. There is nothing contradictory in imagining causal chains that
are closed, though the existence of such chains would lead to rather unfamiliar experiences. For instance, it might then happen that a person
would meet his own former self and have a conversation with him, thus
closing a causal line by the use of sound waves. When this occurs the
first time he would be the younger ego, and when the same occurrence
takes place a second time he would be the older ego.[2] Perhaps the
older ego would find it difficult to convince the younger one of their
identity; but the older ego would recall that an identical experience
once happened to him long ago. And when the younger ego has become
old and experiences such an encounter a second time, he is on the
other side and tries to convince some "third" ego of their physical
identity. Such a situation appears paradoxical to us; but there is nothing illogical in it. However, if such events did occur, there would be
no time order in the usual sense.

Moreover, there would be no unique identity of a physical object in

[2]See Hans Reichenbach, *Philosophie der Raum-Zeit-Lehre* (Berlin and
Leipzig, W. de Gruyter & Co., 1928), p. 167.

time. This *physical identity* of a thing, also called *genidentity*, must be distinguished from logical identity. An event is logically identical with itself; but when we say that different events are states of the same thing, we employ a relation of genidentity holding between these events. A physical thing is thus a series of events; any two events belonging to this series are called genidentical. The relation of genidentity is therefore a two-place propositional function which is symmetrical, transitive, and reflexive.

We apply the relation of genidentity only to events connected by a causal chain. If causal chains are open, it follows that any two different events that are genidentical are not simultaneous. This consequence endows genidentity with a certain similarity to identity; among the events of a temporal cross section, that is, a state of the world given by $t=$ constant, no two different events are genidentical, and genidentity amounts to the same as identity. In other words, if x and y are genidentical events and are simultaneous, then x and y are identical. This consequence, according to which genidentity is *unique* among simultaneous events, appears so familiar to us that we are inclined to regard it as a logical necessity; but it is derivable only for open causal chains, and would be violated in the example of a person meeting his former self.

Fortunately, our world is not of the kind described in this example. When we regard the earlier and later states of a person as states of the same person, this use of the genidentity relation is protected by the law that, at a given time t, a person can exist only in one specimen. But it is not logic which guarantees this law. It is the physical law that there are no closed causal chains which allows us to carry through this definition of physical identity without splitting personalities. These considerations show to what extent the physics of causal chains influences our most fundamental conceptions. Speaking of things and persons that remain identical and unique in the flow of time is possible only because of the causal structure of the universe, because of the openness of the causal net.

Do we have conclusive evidence for the openness of the net? It cannot be said to follow from the equations of Newton's mechanics; it is merely a generalization from experiences in our space-time environment. Yet, for a long time, the openness was never questioned; and it was often regarded as a self-evident presupposition of physics. The general theory of relativity has cast doubts upon this uncritical attitude, because it admits of cosmological structures in which the causal net is closed. Nevertheless, this theory leads to the conclusion that for limited space-time areas the net is open. Here the term "limited" may mean areas of cosmic proportions, and thus we are allowed to regard the causal net of human history, and perhaps of planetary history, as open.

The existence of a linear time order[3] for our physical world is grounded on the openness of the causal net. Assume that a direction has been assigned to the lines; then any two events that are connected by a causal chain and thus are *causally ordered* are also *temporally ordered*. If A lies at the beginning of the chain and B at the end, A is called *earlier* than B, and B is called *later* than A. This determination is unique. If there were another causal chain according to which A was later than B, then these two chains together would constitute a closed causal chain (fig. 6), which case is excluded by the openness of the net.

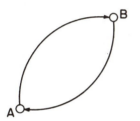

Fig. 6. Two causal lines forming a closed chain. Arrangements of this kind never occur.

There remain events for which no time order is determined, such as the events Q and R in figure 5. If it can be shown that a causal chain which might link these events does not merely happen to be missing, but is physically *impossible*, then such events can be regarded as simultaneous. They may be called *indeterminate as to order of time*.[4]

The time order described is linear, but not yet directed. We can reverse the direction of causally ordered events; then A is called later than B. However, events that are indeterminate as to time order remain in this mutual relation. For this reason, the order of the open net expresses the essential features of our linear time order, except for the direction.

We come thus to an important conclusion concerning statement 3, §2. That the past never comes back is a consequence of the openness of the causal net; for if the past did come back, we would have a closed causal chain. However, the situation would be the same if the direction of time were reversed: what would be the past for this direction would never come back, either. We see that statement 3, §2, does

---

[3]We distinguish between the *lineal* order of the net and the *linear* order of the events on one line. When we speak of the *linear* order of time, we mean that, by a suitable definition of simultaneity, the events on *all* lines can be numbered in corresponding ways; thus these time numbers form a linear series, going from negative infinity to positive infinity.

[4]For this term and the following discussion of simultaneity, cf. Reichenbach, *op. cit.*, §§19 and 22.

not characterize a directional property of time; it merely formulates an order property, namely, the openness of the causal net.

The existence of events between which no causal connection is possible was assumed even in classical physics. For this is the meaning of simultaneity even in the physics of the nineteenth century: simultaneous events are exempt from causal interaction, because a causal influence requires time to spread from one point to another. Introducing the principle that light is the fastest signal, Einstein has shown that the term "simultaneous" is not uniquely defined by the exclusion of causal interaction. Let P and P' be two space points whose world lines are indicated by vertical lines (fig. 7); these lines then indicate the time coördinate. At time t_1 a light signal leaves P; it arrives in P' at time t_2. Reflected at P', it returns to P, where it arrives at time t_3. Since light is the fastest causal chain, *all* events, or time points, occurring at P between t_1 and t_3 are excluded from causal connection with the event occurring at time t_2 in P'. Therefore any one of these events may be called simultaneous with the latter event. This result formulates the *relativity of simultaneity*. It leads to the conclusion that the value of t_2 can be any value between t_1 and t_3, that is, that t_2 is given by the following definition:

$$t_2 = t_1 + \epsilon \left(t_3 - t_1 \right) , \quad 0 < \epsilon < 1 . \quad (3)$$

Any choice of ϵ within the interval between 0 and 1 supplies an admissible definition of simultaneity. The value $\epsilon = 1/2$ offers advantages in certain coördinate systems and is used in the Lorentz transformations for a system in which P and P' are at rest. But this value of ϵ does not define any kind of "true" simultaneity.

These relations are often represented by a diagram like that in figure 8. From the event A, the cone of its possible causes extends into the past, and the cone of its possible effects extends into the future. The spatial volume V_1 contains all the events which, at a

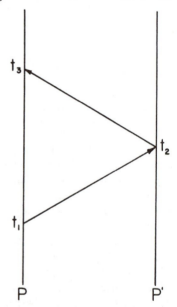

Fig. 7. A light signal leaving point P at time t_1 arrives at point P' at time t_2, where it is reflected; it returns to point P at time t_3. The events of the interval from t_1 to t_3 at point P are excluded from causal interaction with the event occurring at point P' at time t_2.

certain earlier time t_1, can causally contribute to the event A; similarly, the volume V_3 includes all the possible effects of A at the time t_3. To the left and to the right extend the areas of events which cannot enter into causal interaction with A and therefore do not possess a definite time order with respect to A. The present of A is any cross section

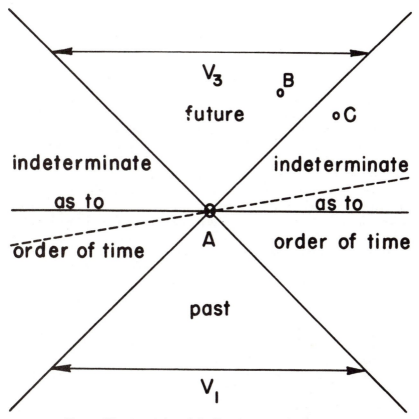

Fig. 8. The time order of the Einstein-Minkowski world.

going through A within these areas, as indicated by the horizontal and the slanted dotted line passing through A. The events situated on one of these lines are assigned the same time values t_2, depending on the value of ϵ used in the definition (3); the horizontal line corresponds to the value $\epsilon = 1/2$. The events A and B determine a *timelike* interval; the events A and C, a *spacelike* interval.

The causal structure assumed for the theory of relativity is the structure of the causal net: it is a structure of time order, not of time direction. Describing the world in negative time, that is, interchanging the cones of past and future in figure 8, makes no difference for the laws that are laid down in the theory of relativity. This is shown by reference to the Lorentz transformations:

$$x = \frac{x' - vt'}{\sqrt{1 - \dfrac{v^2}{c^2}}}, \qquad t = \frac{t' - \dfrac{v}{c^2} x'}{\sqrt{1 - \dfrac{v^2}{c^2}}}. \qquad (4)$$

If in these relations we replace "t" by "$-t$", and "t'" by "$-t'$", we obtain the same equations as before, except that the minus sign in the numerators is changed to a plus sign. This means we obtain the process in which the velocity v has the opposite direction. Reversal of time merely reverses the direction of motion and thus leads to a process governed by the same transformation laws.

It was explained above[5] that the theory of relativity has made a causal definition of time indispensable; this is correct, but the theory does not require a directed time. It is based merely on the assumption of an ordered time. However, its time order would not be permissible if time order were more than causal order. The distinction between timelike and spacelike intervals in the Einstein-Minkowski world is identical with the distinction between causally ordered events and events that are indeterminate as to order of time. For this reason, it remains true that Einstein's theory presupposes a causal definition of time, even if it does not require a definition of time direction.[6]

Since the theory of relativity has not contributed to the problem of time direction, but only to that of time order, it will be permissible, in much of our subsequent discussion, to omit reference to the relativity of simultaneity. For the sake of convenience, we shall therefore often speak of the present as though it were a determined cross section through time. In precise parlance, we would have to say in such situations that, among the many admissible cross sections, we can select one which we regard as the present. For distances of the dimensions of the earth's surface, this ambiguity in the definition of the present is negligible.

---

[5] See p. 39.

[6] The difference between an order and a direction of time was explained in Reichenbach, *op. cit.*, pp. 164–165, where it was pointed out that the theory of relativity presupposes merely an order of time, not a direction.

● 6. Intervention

Thus far we have defined an order of time. With the intention of proceeding to a definition of time direction, we might suppose that it may help us to study processes in which we change the future through acts of *intervention*. Since we cannot thus change the past, it seems as though we possess here a way of discriminating between past and future.

Our daily life abounds in suitable illustrations. We turn on a switch, and a lamp starts burning. If we had not turned the switch, the lamp would have remained in its preceding state, in which it did not emit light. We hit an oncoming tennis ball with a racket, and the ball is returned; if we had not hit the ball, it would have gone on in the original direction. We light a pile of paper with a match; if we had not done so, the pile would not have burned to ashes. We insert a red glass into a beam of white light and thus transform the later part of the beam into red light, whereas the earlier part remains unchanged; if we had not inserted the glass, the beam would have been white all the way. In all these examples, we perform acts of intervention. The fact that intervention is possible, and occurs often, seems to indicate clearly that we can change the future, but cannot change the past.

From the illustrations we see that this claim to have a power over future happenings is based on a *conditional contrary to fact*. We make a statement about what would have happened if we had not intervened. We therefore have to study these counterfactual conditionals more precisely if we wish to understand what we mean by changing the future.[1]

The last two examples refer to irreversible processes, since the absorption of light in a red glass and the burning of paper into ashes are irreversible occurrences. Since we know that such processes can define a time direction, let us postpone their discussion to a later chapter. Let us study the nature of intervention by reference to reversible processes.

Assume that a tennis ball travels from A to B (fig. 9), where it is hit by a tennis racket in such a way as to move on in the direction toward C. We say that the intervention performed by means of the racket has changed the path of the ball; that if there had been no intervention, the ball would have gone on beyond B in the direction determined by the line AB. How can we prove this conditional statement? The only way to prove a conditional contrary to fact is to carry through the process under varied conditions. We let the ball go from A to B without

---

[1] For a detailed discussion, see Hans Reichenbach, *Nomological Statements and Admissible Operations* (Amsterdam, North-Holland Publ. Co., 1954).

putting a racket in its path; the ball then travels toward D (fig. 10).
This simple variation of the process by omission of the use of the
racket seems to prove our conditional contrary to fact.

On closer inspection, however, we find that we have used a tacit
presupposition. We have constructed an experiment which was deliber-
ately so arranged as to leave the path AB unchanged; therefore, only
the second part of the ball's path was changed. Can we construct a
different experiment in which only the second part of the path, going
from B to C, remains unchanged? This is easily done (see fig. 11). We
throw the ball from E toward B and do not put a racket into its path;

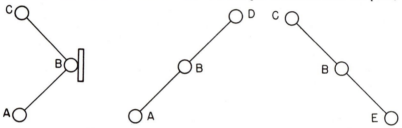

Fig. 9 (left). Path of a tennis ball reflected by a tennis racket.
Fig. 10 (center) and fig. 11 (right). Different paths of a tennis ball without
reflection at a tennis racket.

then the ball goes on toward C. Here we have performed a second kind
of variation by the omission of the use of the racket; but in this experiment the second part of the ball's path remains unchanged, while the
first part is changed.

We shall argue: The last experiment cannot be used to study intervention. Here we have produced the same effect, namely, the traveling
of the ball from B to C, by an *alternative cause*. Therefore, this experiment merely proves the existence of an alternative cause.

This consideration, however, leads to an important conclusion. The
process shown in figure 9, taken alone, does not tell us whether the
tennis racket functions as an intervention or as an alternative cause.
These two terms refer to relations between two experiments: when we
compare the process represented in figure 9 with that diagramed in
figure 10, we say that the tennis racket performs an act of intervention;
when we compare the process shown in figure 9 with that shown in
figure 11, we say that the tennis racket acts as an alternative cause.

We see, furthermore, that the question "What would have happened
if the tennis racket had not been inserted?" cannot be answered unless
a further specification is given. Either one of the processes of figure
10 or figure 11 could then have happened. The conditional contrary to
fact "If the tennis racket had not hit the ball, then . . ." is ambigu-

ous; the blank in it cannot be filled in before we say whether we want to keep the part AB or the part BC of the ball's path unchanged. But seen from the event B, these parts represent the past and the future, respectively. When we say that hitting the ball with the racket changes the future path of the ball, our statement presupposes the tacit antecedent, "if we assume the past to remain unchanged". No wonder that acts of intervention change only the future, and do not change the past; the term "intervention" is defined by the condition that the past be unchanged. The statement that acts of intervention cannot change the past is a trivial tautology.

This consideration leads to the conclusion that acts of intervention cannot define a direction of time. The term "intervention" is defined only after a direction of time is given; acts of intervention are actions that leave the past unchanged.

This result is made even more obvious when we describe the processes in negative time. The terms "intervention" and "alternative cause" then interchange their roles. What is called an intervention in positive time becomes an alternative cause in negative time, and vice versa. This is illustrated by figures 9–11. If figure 9 is interpreted in negative time, the ball goes from C by way of B to A. Comparing figure 9 with figure 11, we now call the hitting by the racket an intervention; and comparing figure 9 with figure 10, the hitting by the racket is called an alternative cause. In negative time, too, we would say that we cannot change the past, but can change the future: the intervention by the racket makes the ball travel from B to A instead of going from B to E.

This result leads to an interesting criticism of statement 4, §2, according to which we can change the future, but cannot change the past. As long as reversible processes are used for intervention, this statement, like statement 3, does not express a direction of time. If the term "past" is defined in terms of a given time direction, and thus in a *relative* sense, statement 4 is true as a trivial tautology, and the statement holds for both time directions. However, what we call the past in positive time can be changed in negative time, because for negative time, it is future. Consequently, if the word "past" is restricted to denote that stretch of time which it denotes in positive time, and thus is used in an absolute sense, we can change the past by using negative time. In this sense, therefore, statement 4 is false.

There is a third sense in which we can interpret statement 4. We can regard it as meaning that going on along a causal chain we cannot influence points reached before, because such influence would constitute a closed causal chain. In this interpretation, statement 4 is true; once more, however, it is not a directional but an order statement,

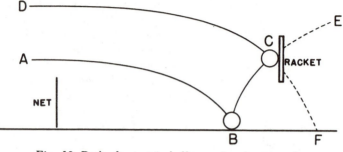

Fig. 12. Path of a tennis ball crossing the net and
being returned by a tennis racket.

because it is true whether we proceed in positive or negative direction
along a causal chain.

To illustrate these results, let us describe the happenings in a
tennis game in negative time. Assume that we take a film of a tennis
game and then reel the film in reverse. The ball goes back and forth
as in the original game, the only difference being that every ball is
received as a volley, that is, it does not touch the receiver's court
before it hits his racket; on his return of the ball it may or may not
touch his court before crossing the net. The movement of the racket
appears as the cause of the return of the ball; we would say that if the
racket had missed the ball, the ball would have gone on in the same
general direction. This statement, however, has a meaning different
from the one it has for positive time. Let the diagram of figure 12
represent the path of the ball in a vertical plane; in positive time, the
ball travels in the direction *ABCD*; in negative time, in the direction
DCBA. For positive time, we say: Had the racket missed the ball, the
ball would have traveled from *C* to *E*. For negative time, in contrast,
we say: Had the racket missed the ball, the ball would have traveled
from *C* to *F*. In both interpretations, the intervention of the racket
changes that part of the ball's path which we call the future.

A player's service would look rather strange in reverse time. The
other player sends him a fast ball, which bounces off the ground before it crosses the net. The serving player moves his racket backward until it is hit by the ball. The impact makes the player's arm
swing all the way back. The ball, bouncing off the racket, travels
straight up, and on falling down is caught by the player's left hand.
Though an unfamiliar occurrence, the motion of the ball is governed by
the laws of mechanics in the same way as in the usual interpretation.
The mechanics of tennis balls does not tell us that the film was reeled
backward.

It would be a strange experience indeed to see the players run backward. Such a motion, although compatible with the laws of mechanics, is unusual because we are safer if our steps are controlled by our eyes. This problem, however, requires the discussion of irreversible processes, which we shall treat in a later chapter.

CHAPTER III THE TIME DIRECTION OF
THERMODYNAMICS AND
MICROSTATISTICS

• 7. Report on the Second Law of Thermodynamics

Although irreversible processes belong to the familiar experiences of everyday life—we may recall such processes as burning, the growth of living organisms, the mixture of gases or liquids—they were not treated scientifically until the nineteenth century. On the basis of Carnot's discoveries (1824) of the laws governing thermal processes of cyclical form, also called Carnot cycles, R. E. Clausius (1850) and W. Thomson (1851, later Lord Kelvin) formulated the second law of thermodynamics, a principle that makes possible the mathematical expression of a direction controlling the course of physical occurrences. By its name, this principle was characterized as a parallel of the first law of thermodynamics, better known under the name of the law of conservation of energy. The two laws together determine the ways in which physical quantities change during processes involving the phenomenon of heat. The first law states that in all changes there exists a certain quantity, called *energy,* which retains a constant value. In its classical form the second law states that there exists another quantity, called *entropy,* which in some changes remains constant, but in other changes increases, whereas it is impossible that this quantity ever decrease. Irreversible processes are those in which entropy increases.

This formulation makes the parallelism of the two principles obvious; but it also explains why only the second law allows us to infer a direction for physical processes. If a state A is transformed into a state B and the entropy remains constant in this transition, then it will remain constant also in the inverse transition from B to A. If, however, during the transition from A to B the entropy is increased, that is, if the entropy in state B is larger than the entropy in state A, then the inverse transition would represent a decrease in entropy; and if it is a physical law that entropy never becomes smaller, the reverse process is excluded. The existence of a direction for physical processes is thus formulated by means of a state function S, the entropy, which has a determinate numerical value for every given state and orders physical states in terms of increasing numbers.

In the formulation of these principles, of course, care has to be taken that all processes referred to are closed, that is to say, that all changes involved are included in the system considered. Energy can flow from one system to another; then one system loses energy and the other gains it. But the sum total of losses and gains is zero. Likewise,   ·
entropy can flow from one system to another; then one system loses entropy and another gains it. But only for reversible processes is the sum total of all changes in entropy equal to zero; for irreversible processes, it is always some positive amount. That is the assertion of the second law of thermodynamics.

Entropy is a quantity defined in terms of thermodynamical parameters, such as temperature and pressure. Mechanical parameters—for example, position and velocity—do not enter into the definition of its value; that is, there is no such thing as a mechanical entropy. This conception can be carried through for the following reason. If a system goes through a purely mechanical process, that is, a process during which its temperature, heat content, and so forth, do not change, the process is reversible; therefore its entropy must remain constant. In other words, a change in mechanical parameters has no influence on the value of the entropy. Only when, in addition to mechanical changes, there are thermodynamical changes involved, such as heat generation through friction, will the entropy of the system change. The reversibility of mechanical processes makes the concept of a mechanical entropy dispensable.

However, though most thermodynamical processes are irreversible, there also exist some reversible processes which involve changes in heat conditions. Such processes, at least, can be constructed as limiting cases of irreversible processes. If the temperature difference between two systems is very small and the systems come into interaction, heat will flow very slowly from the warmer to the colder system, and the process is one in which the system goes through a continuous series of states which can be regarded as states of equilibrium. For such processes, one system gains in entropy as much as the other system loses, and the sum total of entropy exchanges is zero.

This consideration leads to two meanings of the word "reversible". In the narrower sense, it applies to the total system, which is closed and whose entropy remains constant. In this sense, reversibility is defined only for closed systems: a process occurring in a closed system is reversible if it can also proceed in the opposite direction while the system remains closed. In the wider sense, the process undergone by a part of such a system is also called reversible. Used in this sense, the term "reversibility" can apply to open systems, and we arrive at the paradoxical formulation that a system can either increase or decrease its entropy in a reversible process. The paradox, which has puzzled many a reader of textbooks on thermodynamics, is easily solved when it is recognized that here the word "reversible" is used in the wider sense. It appears advisable, however, to avoid the wider meaning of the term, and to speak of reversibility only for the total process gone through by a closed system.

A reversible process involving changes in heat conditions is represented by the Carnot cycle. If this process goes in one direction, heat is transferred from a warmer to a cooler reservoir by way of a gas system which moves a piston in a cylinder and produces mechanical energy. If the process is run in the opposite direction, mechanical energy is consumed, and heat is transmitted from a cooler to a warmer reservoir. In the first form, the process is used for steam engines, in which the reservoirs consist in the boiler and the cooler of the machine; in the second form, it represents a refrigeration process. The entropy of the total system, including the reservoirs and the gas system, remains constant throughout the cycle; that of its parts changes periodically in such a way that when the gas system gains entropy one of the reservoirs loses it, and vice versa.

Strictly speaking, reversible processes never occur: if we wish to run the Carnot cycle in reverse, we have to feed a little more mechanical energy into it than is gained while the cycle is run in the first

direction. But the slower the process, the smaller the difference in
the two amounts of mechanical energy. This is what is meant when we
say that reversible processes can be defined as limiting cases of
actual processes. The same is true, incidentally, for the reversible
processes of mechanics. In all actual mechanical processes there is
some heat generation through friction; completely reversible processes
do not exist.

If the entropy S changes during a process which is part of a reversible process (that is, a partial process that is reversible in the wider
sense), and if dQ is the amount of heat entering or leaving the system,
and T the absolute temperature, the change satisfies the equation

$$dS = \frac{dQ}{T} \ . \tag{1}$$

Here $dS \gtrless 0$, depending on whether $dQ \gtrless 0$. If $dS > 0$ for the process
considered, we have $dS < 0$ for the remainder of the total process. It
can be shown that the integral of (1) is independent of the path and
that the entropy S therefore has the properties of a thermodynamical
potential. This fact can be used to compute entropy, or rather, entropy
differences, for processes which are not parts of reversible processes.

If we throw an ice cube into a thermos bottle containing warm water,
we start a process of heat equalization in a closed system, a process
which is neither reversible nor part of some reversible process. The
entropy of the total system increases slowly. Here relation (1) is not
satisfied by the total system, because $dS > 0$ and $dQ = 0$. However, it
is possible to connect the initial state and the final state by some
other process which is not closed and which is part of a reversible
process. For this other process, formula (1) is satisfied, and we can
thus compute the entropy difference by integration of dS. By way of
this detour, entropy values, to within an additive constant, can be
computed for irreversible processes which are not parts of reversible
processes. All irreversible processes going on in closed systems
belong to this category.

By considerations of this kind, R. E. Clausius and W. Thomson have
computed the value of the entropy S in terms of thermodynamical parameters. For instance, a volume v of a gas whose mass amounts to one
unit, has the entropy (up to an additive constant)

$$S = c_v \log T + \frac{R}{m} \log v \ , \tag{2}$$

where c_v is the specific heat at constant volume, T the absolute temperature, m the molecular weight of the gas, and R the general gas

constant. Given the values of these parameters, we can compute S.
If two systems are brought together, their entropies are additive:

$$S = S_1 + S_2 \ . \tag{3}$$

Consider, for instance, two equal quantities of the same gas which have different temperatures and are brought into thermal interaction by being put into adjacent containers. Then heat flows through the walls from the gas of higher temperature to that of lower temperature. According to (2), the entropy of the first gas will then decrease, and that of the second will increase; this follows because formula (2) contains the temperature within the logarithmic function, which increases with its argument. The temperature of both systems will finally assume a mean value; but the logarithm of the mean is larger than the mean of the logarithms, and the sum (3), taken after the temperature is compensated, is larger than the corresponding sum at the beginning.

To consider another illustration, assume that a gas is in a container connected with another container by a window. We close the window and evacuate the second container. If now the window is opened, the gas flows into the vacuum, thus filling a larger volume. After equilibrium has been reached, the temperature is the same as before, because no external work has been done. From equation (2) we see that then the entropy S increases, the quantity v becoming larger. This means that mere expansion of a gas into a larger volume is an irreversible process. If we want to bring the gas back to its original volume, which of course we can do, we have to intervene from the outside; that is, the system is no longer closed.

Since completely reversible processes do not occur, the second law of thermodynamics can be stated in the form that the entropy of a closed system increases as long as any processes are going on within it, or, in other words, so long as a state of equilibrium has not been reached. However, this does not mean that every process is directed. If a system is not closed, the direction of the process depends on the interaction of the system with other systems. We cannot say that heat always flows from the warmer to the cooler body. The inverse Carnot cycle illustrates the contrary, and so does every refrigerator. But the refrigerator can cool objects only because at the same time it transforms electrical energy into heat of room temperature through the action of its pump, which uses electric current and ejects the used energy through a radiator into the room. Electricity represents a highly ordered form of energy. Its transformation into heat produces an increase in entropy which exceeds the decrease resulting from the cooling inside the box. The total system, therefore, gains entropy, while, as a causal consequence, one of its subsystems loses entropy. And the principle that in closed systems the entropy cannot decrease is satisfied.

These simple illustrations show that the existence of a direction for physical processes can be expressed by the use of a state function, the entropy, which, so to speak, measures the *degree of equalization* reached by a system. All processes are directed toward increasing equalization. This result applies not only to individual processes, but also to the universe as a whole.

Let us assume that an instantaneous cross section through the universe, that is, one of its momentary states, represents a physical state which can be assigned a definite value of entropy. The discussion of objections to this assumption will be postponed to a later section (see p. 116), where it will be shown that the present argument does not depend upon this assumption. If the universe as a whole possesses at every moment a specific entropy, this value is subject to the general law of entropy increase; this means that the universe progresses toward more and more equalized states. Although this principle leads to the unwelcome consequence that someday our universe will be completely run down and offer no further possibilities of existence to such unequalized systems as living organisms, it at least supplies us with a direction of time: positive time is the direction toward higher entropy.

To these dreams of classical physics a discovery by the Viennese physicist Ludwig Boltzmann (1872) put a sudden end. He found that the principle of the increase of entropy must be regarded not as a strict, but as a statistical, law. We should not say, "Entropy must become larger", in the way that we say, "Energy must remain constant". We should say, "It is highly probable that entropy will become larger". In other words, the reverse process of a thermodynamical occurrence is not physically impossible; it is merely very improbable.

Using the results of the kinetic theory of gases, according to which the heat content of a gas consists in the irregular mechanical motion of its molecules, Boltzmann showed that the collisions of molecules are governed by statistical laws which lead to an average equalization of differences in speed. When a fast molecule hits a slow one, it may occasionally happen that the slow molecule loses most of its speed and imparts it to the fast one, which then travels away even faster; but such occurrences are exceptions. In the overwhelming number of collisions, the faster molecule will lose speed and the slower one will gain it. The passage of heat from higher to lower temperature is thus to be understood as a statistical equalization of differences in molecular speed. The law of the increase of entropy is guaranteed by the law of large numbers, familiar from statistics of all kinds; but it is not of the type of the strict laws of physics, such as the laws of mechanics, which are regarded as exempt from possible exceptions.

Probability laws had been applied to arrangements of molecules in earlier phases of the evolution of the kinetic theory; in particular, J. C. Maxwell had shown that the speeds of the individual molecules are distributed according to a normal curve. However, Boltzmann recognized that the probability W of a state is related to that state's entropy S, and he established the famous relation (holding to within an additive constant):

$$S = k \cdot \log W \ . \tag{4}$$

That the entropy increases during physical processes means, in this interpretation, that ordered arrangements of molecules are transformed into unordered ones. It is easily seen that the probability of the latter states exceeds by far that of the former; order is an exception in a world of chance. That entropies are additive, according to (3), follows because probabilities are multiplicative; the logarithmic function of (4) transforms the multiplication into an addition. This is more than a qualitative correspondence; formulas like (2) are strictly derived in Boltzmann's gas theory, which thus achieves a complete reduction of thermodynamics to statistical mechanics.

The direction of physical processes, and with it the direction of time, is thus explained as a statistical trend: the act of becoming is the transition from improbable to probable configurations of molecules. It is this interpretation of time direction which we shall have to study in the following sections. We shall see that it represents, in fact, the nucleus of a theory of the flow of time.

Another important result of Boltzmann's interpretations may be discussed in this connection. It concerns the aforementioned distinction between the two kinds of physical law. Both kinds of law are expressed by implications, that is, they are if-then statements. But whereas the strict, or causal, law expresses a *strict implication,* or *nomological implication,* the probability law expresses merely a *probability implication.* The first has the form, "If . . . , then always . . ."; but the second has the form, "If . . . , then in a certain percentage of cases . . .". If the probability is low, the statistical character of the law is apparent; but if it is high, one might easily mistake a probability law for a strict law. This had been done, in fact, for the second law of thermodynamics; and it is Boltzmann's great discovery that this law belongs to the statistical category.

With the recognition of the distinction between the two kinds of law there arose an important question. Which of the two kinds is more fundamental? Is it possible to derive the one from the other? In Boltzmann's time, it was the general opinion that the strict type of law has a logical priority. The use of statistical considerations in thermodynamics appeared to be merely an expedient, forced upon us by the limitations of man's abilities; it was assumed that if we could only observe every molecule in its path, we would be able, like Laplace's superman, to predict strictly the future development of the gas, and we would not have to resort to statistical considerations. Statistical laws thus appeared to be laws *faute de mieux*—formulations tentatively adopted for want of something better. According to this view, statistical laws are ultimately reducible to causal laws. We shall discuss this problem in §§10 and 11.

By contrast, the opposite opinion was uttered that perhaps statistical law is the fundamental type of law. We have no proof that the motion of atoms is governed by strict laws. Perhaps we should explain all causal laws of macrophysics as the product of the law of large numbers, which transforms the limited probability of elementary occurrences into the high probability of processes in large assemblages. The fate that had befallen the second law of thermodynamics then would be the prototype of a similar interpretation for all other so-called strict laws of physics.

The development of modern quantum physics has decided in favor of the second opinion. However, the discussion of this conception will be postponed to chapter v. We must analyze the possible interpretations of classical physics before we can understand the turn made in quantum physics. And we must investigate whether it is true that classical physics can do without the concept of probability. We shall see that classical physics is in need of this concept as much as quantum physics is, although it employs the concept in a different form.

• 8. The Statistical Definition of Entropy

Boltzmann's gas statistics, in its original form, is a combination of causal and probability methods. He assumes that all elementary processes are determined by strict causal laws, and he wishes to show that the totality of such processes is governed by statistical laws. In order to derive this conclusion, he introduces a certain probability assumption, which is to be added to the causal laws of physics.

The methods and assumptions used for these considerations may be illustrated by a very simple example. Let us study the distribution of a gas filling a container. We wish to prove that an even distribution over the space accessible to the gas is the most probable one, that is, that its entropy is at a maximum.

We divide the container into a number of cubical cells of equal size; let m be the number of the cells. A *distribution* is given when we know the number n_i of molecules within every cell. In an uneven distribution —for instance, while the container is being filled with gas—these *occupation numbers* n_i may be very different. Through the movement of the molecules and their collisions a certain shuffling takes place, and the development toward an equal distribution may be regarded as a transition from an improbable to a highly probable distribution. In the state of equilibrium, the occupation numbers n_i will be approximately equal and remain so, although the individual molecules continually travel from cell to cell. The sum of all n_i must always be equal to the number n of all molecules in the container:

$$\sum_{i=1}^{m} n_i = n \ . \tag{1}$$

A distribution of molecules must be distinguished from an *arrangement*. For the definition of "arrangement", we assume that every molecule is an individual entity, distinguishable, in principle, from every other one, and that we know which cell it occupies. *A distribution is a class of arrangements*. For instance, if a certain molecule in the cell i and another one in a different cell k exchange their places, we obtain a different arrangement, or "complexion", whereas the distribution remains the same. However, if two molecules in the same cell i exchange their places, even the arrangement remains unchanged, since the size of the cell defines the limit of exactness within which we wish to specify an arrangement.

In order to compute the probability of an arrangement, Boltzmann introduces the assumption that it is equally probable for a molecule to occupy any one cell. This is the *probability metric* on which the computation is based; it can be written in the form

$$\varphi\left(x_1, \ x_2, \ x_3\right) = \text{const} \ , \tag{2}$$

where x_1, x_2, x_3 are the spatial coördinates and $\varphi(dx_1, dx_2, dx_3)$ is the probability of finding a specified molecule in the volume element $dx_1 dx_2 dx_3$. Since the number of cells is m, this means that the probability of finding a certain molecule in a certain cell equals $1/m$. The probability of any given arrangement therefore equals $(1/m)^n$, if it is assumed that an arrangement can be considered as a conjunction of n independent events, each of probability $1/m$. Vice versa, the total number N of arrangements is given by

$$N = m^n \; . \tag{3}$$

This value is easily computed, as follows. One molecule can occupy any of the m cells, and thus makes m arrangements possible; adding a second molecule, we see that each of its m possible arrangements can be combined with each of the preceding arrangements, so as to furnish m^2 arrangements, and so on.

Since a distribution is a class D of arrangements, the probability of a distribution is computed by counting the number N_D of arrangements belonging to it. This method involves the assumption mentioned, namely, that the molecules are distinguishable individuals, and consequently that the interchange of two molecules in different cells leads to two different arrangements. The probability of a distribution D is then given by

$$W = \frac{N_D}{N} = N_D \cdot \frac{1}{m^n} \; . \tag{4}$$

The number N_D is found by permuting all the molecules in the cells, while the numbers n_i remain constant. Permutations of this kind are considered in Bernoulli's theorem,[1] one of the fundamental theorems of the calculus of probability; it is shown there that the number of such permutations is given by

$$N_D = \frac{n!}{n_1! \cdots n_m!} \; . \tag{5}$$

It is convenient to compute $\log N_D$ rather than N_D itself. Furthermore, it is sufficient to regard only cases in which the n_i are not too small; we can then use Stirling's famous approximation for factorials,

$$\log n! = n \log n - n + \tfrac{1}{2} \log 2\pi n \; . \tag{6}$$

For large n, the third term is small as compared with the other two terms; that is, if it is divided by n, the quotient goes to 0 with growing n. Now we will assume not only n but also the various n_i to be large, neglecting extreme forms of an uneven distribution. We then can omit the third term. Applying the remaining expression to n and to each of the n_i, we find

$$\log N_D = n \, \log n - \sum_{i=1}^{m} n_i \, \log n_i \; . \tag{7}$$

[1]See, for instance, Hans Reichenbach, *The Theory of Probability* (Berkeley, Calif., Univ. Calif. Press, 1949), p. 264. This book will be cited hereafter as *ThP*.

These terms result from the first term in (6), when (6) is applied to the numerator and denominator of (5). The second term of (6) drops out because of (1).

Using this convenient simplification for $\log N_D$, we can now proceed to computing $\log W$. Combining (7) with (4) and (3), we find, using Boltzmann's identification of entropy and the logarithm of state probability,

$$S = \log W = n \log n - \sum_{i=1}^{m} n_i \log n_i - n \log m \ . \tag{8}$$

This expression represents a *normalized* form of the entropy, in the sense that the expression contains no undetermined constants. Furthermore, the form (8) is normalized in the sense of a probability, since it results directly from the logarithm of the probability of the gas state, though with the qualification that the last term in Stirling's formula (6) has been omitted. But since this term is small, relation (8) can be regarded as a good approximation. It defines entropy as a function of occupation numbers; and the replacement of a probability by its logarithm ensures the additivity of entropy.

Boltzmann writes this relation somewhat differently. In order to arrive at a correspondence with the value S of Clausius' and Thomson's entropy, he introduces the constant k of relation (7, 4); furthermore, he renounces the explicit formulation of the last term in (8), replacing it by an undetermined constant and thus making different interpretations of it possible. He therefore writes the value of the entropy in the *non-normalized* form:

$$S = k \left[n \log n - \sum_{i=1}^{m} n_i \log n_i \right] + \text{const} \ . \tag{9}$$

It is convenient to study a third form of entropy, which, like (8), contains no undetermined constant, but differs from (8) by a factor. Let us introduce the quotient

$$p_i = \frac{n_i}{n} \ . \tag{10}$$

The value p_i can be regarded as the conditional probability of finding a molecule in the cell i, on the condition that the gas possesses the distribution D considered. Dividing (8) by n, we have

$$S = \frac{1}{n} \log W = \log n \left(\sum_{i=1}^{m} p_i \right) - \sum_{i=1}^{m} p_i \, \log n_i - \log m$$

$$= - \sum_{i=1}^{m} \left[p_i \, \log n_i - p_i \, \log n \right] - \log m \tag{11}$$

$$= - \sum_{i=1}^{m} p_i \, \log p_i - \log m \; ,$$

remembering that

$$\sum_{i=1}^{m} p_i = 1 \; . \tag{12}$$

The value S of (11) differs from that of the form (8) by the factor $1/n$ and may be called the *specific entropy*, since it is independent of the number n of molecules. The values S of both (8) and (11) are normalized in the sense of probabilities; they are thus negative numbers, because W is a fraction, $W \leq 1$, and its logarithm ≤ 0.

Compared with the original entropy of nonstatistical thermodynamics, the *statistical forms* (8), (9), and (11) represent a certain generalization. In nonstatistical thermodynamics, entropy is defined only for states of equilibrium. In contrast, the statistical forms define a value of S for all kinds of conditions, such as exist when the gas molecules are distributed unevenly over the space. Assigning to states of nonequilibrium a statistical form of entropy provides a means of explaining the movement toward equilibrium: states of equilibrium possess a higher probability, and the evolution of the gas from states of uneven distribution to states displaying equalization appears as a tendency to proceed from states of lower to states of higher probability.

We will now study the problem of determining the distribution which represents the state of equilibrium, that is, the distribution which corresponds to the maximum of W. Since the factor $1/N$ in (4) is constant, this distribution is found by maximizing (5). It is convenient to select the form (7) for the maximization, because this form makes use of the simplification introduced by Stirling's formula. The determination of the values n_i corresponding to the maximum of (7) is somewhat complicated, since the n_i are subject to condition (1). For a maximization under a certain restrictive condition, Lagrange has developed a method which uses infinitesimal variations in combination with a *multiplier* α and which leads to the result that

$$\sum_{i=1}^{m} \left[\delta \left(n_i \, \log n_i \right) + \alpha \cdot \delta \left(n_i \right) \right] = 0 \; . \tag{13}$$

The first part stems from the variation of (7), the second from the variation of (1). When we carry out the differentiation, the Lagrange factor α allows us to regard the δn_i as independent quantities; thus each term of the sum must vanish, and we arrive at the relation

$$\log n_i + 1 + \alpha = 0 \ ,$$

$$n_i = e^{-1-\alpha} = \text{const} = \frac{n}{m} = q \ . \tag{14}$$

The value of the constant follows from (1); we call it q. This result means that maximum probability is associated with the case that all cells are occupied by equal numbers of molecules. Combining (14) with (10) we see that, for this case, all the p_i are equal, and, in particular, they each equal $1/m$. This, then, is the condition that the entropy S reaches its maximum.

Whereas the maximal value of Boltzmann's entropy S still depends on the value of the constant in (9), it is easily seen that the normalized maximum entropies (8) and (11) assume the value $S = 0$. This follows for (8) when we replace "n_i" by "n/m"; for (11), when we replace "p_i" by "$1/m$". For the probability W this means that here $W = 1$. This result is not strictly correct; the probability W can only be very close to 1. The incorrectness results from neglecting the last term in Stirling's formula (6). But, for all practical purposes, we can assign to the state of equal distribution of the molecules the probability 1.

In his writings, Boltzmann uses, instead of S, a function H which is the negative of S and is therefore given, in application to the specific entropy (11), by the form:

$$H = \sum_{i=1}^{m} p_i \log p_i + \log m \ . \tag{15}$$

For gas theory, this is merely a more convenient notation, because $H > 0$. It leads to the consequence that H tends toward a minimum, whereas S goes toward a maximum. We shall see later that H can be given a particular significance in a different field of application, namely, in information theory.

Let us study in more detail how the statistical interpretation of entropy describes the transition to states of equilibrium. We will consider the simple example of a gas spreading from one container into an evacuated container of equal size, after a window between the containers has been opened. Using the form of entropy given in (8), which is normalized in the sense of a probability, we may describe this process as follows. As long as the window is closed, the gas is in a space of m cells in a state of equilibrium. The occupation numbers n_i thus have the constant value n/m given in (14). When we denote by S_1 the entropy for this state, it is easily seen that (8) yields the value $S_1 = 0$. The probability of the gas state is thus equal to 1. When we now open the window, the number of cells is raised to $2m$; but in the beginning the new cells are all empty. The summation term in (8), therefore, though now extending to $2m$, yields the same value as before, whereas the third term of (8), in which "m" is replaced by "$2m$", yields an added term "$-n \log 2$". Immediately with the opening of the window, the entropy thus decreases to a value S' given by

$$S' = S_1 - n \log 2 . \tag{16}$$

Since $S_1 = 0$, we have here a probability of $W = 1/2^n$. This low value of the probability results because a probability is always relative; and with the increase in accessible cells, the reference class of the probability has changed. But the low probability cannot be maintained; the gas pours into the empty container, filling its cells. In the final state of equilibrium, each of the $2m$ cells is occupied by $n/2m$ molecules. When we compute the value (8) by putting $n/2m$ for n_i and extending the summation to $2m$, we find again the value $S = 0$. The gas thus comes to its equilibrium with a value S_2 of its entropy such that $S_2 = S_1$.

This is the probability interpretation of the process; but it differs in one essential point from the thermodynamical interpretation. The thermodynamical entropy is a state function which does not change when the window of the container is opened; in fact, the regress of the entropy expressed in (16) appears here quite impossible. The gas keeps its entropy S_1 when the window is opened; but with the beginning of the spreading process the entropy increases, and when finally the second container is filled, the total system of the two containers has reached an entropy S_2 such that $S_2 > S_1$.

This consideration shows that the thermodynamical entropy cannot be identified with the S of the form (8), which is normalized in the sense of a probability. It requires a function which does not change its value when the window between the containers is opened, but increases when the gas spreads into the empty container. This means that thermodynamics requires an *absolute entropy* and cannot be interpreted by the use of a *relative entropy* stemming from adjustment to a probability. A suitable quantity is found when we identify S not with $\log W$, but with $\log N_D$, that is, when we measure entropy by the absolute number of arrangements rather than by their relative number. Formula (7) tells us that in the definition of S we would then omit the third term in (8). We also could use a form containing an additive constant if we interpret this constant as not depending on m, but merely being of the form "$n \cdot$ constant". The dependence on n is necessary in order that the entropy be doubled when the number n of molecules is doubled. In this interpretation, Boltzmann's entropy (9) is an absolute entropy.[2]

Although the use of an absolute entropy appears unobjectionable at first sight, it leads to certain difficulties, which were first pointed out by Gibbs. Suppose we have again two containers side by side, but this time the second container is filled by a gas in exactly the same physical condition as the first. As long as the window between the containers is closed, each gas system has the entropy S_1, and the total system has therefore the entropy $2S_1$. When we now open the window, nothing happens, macroscopically speaking; in fact, the total gas system is in equilibrium. Its entropy S should therefore still be given by the value $2S_1$. Now it is easily seen that when we use the relative entropy (8) this requirement is satisfied. If in (8) we replace "n" by "$2n$" and "m" by "$2m$", while the values n_i remain unchanged, the value of the whole expression is simply doubled. An extra term "$2n \log 2$" arising from the first term is here compensated by a term "$-2n \log 2$" stemming from the third term. However, when we omit the third term and use an absolute entropy, we find the relation

$$S = 2S_1 + 2n \log 2 \ . \tag{17}$$

Opening the window has increased the entropy, although no macroscopic change occurs for the gas; and the entropy of the total system is larger than the sum of the entropies of the two individual systems.

This *paradox of normalization*, also called Gibbs's paradox, is inherent in the use of an absolute entropy for Maxwell-Boltzmann statistics.[3] A satisfactory interpretation of thermodynamical entropy in terms of the absolute number of arrangements is here not possible, because an absolute entropy cannot be consistently normalized. The entropy of the whole is not the sum of the entropies of its parts; thus the *additivity postulate* is violated. The physicist helps himself by using what may be called a probability interpretation *by stretches*. If a thermodynamical system goes from the entropy S_1 to the entropy S_2, and then from S_2 to S_3, the physicist coördinates a probability increase to the transition from S_1 to S_2 and another probability increase to the transition from S_2 to S_3; but these probability increases are not conjoined

---

[2]For a somewhat different normalization of the constant of an absolute entropy, see §9, n. 1, of the present study.

[3]The paradox would disappear if we were to omit in (8) not only the third term, but also the first term. However, another paradox would then arise when the number n of molecules in a container is doubled. Instead of going from S to $2S$, which is the required value, the entropy would go from S to $2S - 2n \log 2$.

into one increase. There has to be a repeated regauging: first, S_1 is replaced by a low value S_1', from which the system proceeds to a high value S_2' corresponding to S_2; secondly, S_2' is replaced by a low value S_2'', from which the system proceeds to a high value S_3'' corresponding to S_3; and so on. In the mathematical technique, this regauging of probabilities is expressed in the occurrence of undetermined constants; entropy formulas are true, subject to suitable adjustment of an additive constant. This *additive openness* is a general feature of the entropy of Maxwell-Boltzmann statistics. It will be seen that quantum statistics is superior in this respect and need not resort to entropy relations which are additively open.

It was explained above that the relative forms (8) and (11) of the entropy are not subject to Gibbs's paradox. The satisfaction of the additivity postulate was proved by direct computation. It should be realized, however, that these computations are only approximately true, because they neglect the last term in Stirling's formula (6). If this term is introduced into the computations, we arrive here, too, at contradictions to the additivity postulate. For instance, we find that, if $W(n, m)$ is the probability of a distribution of n particles in m cells, given by (4) and (5), the relation

$$W(2n, 2m) = W(n, m)^2 \qquad (18)$$

is not strictly true. The left-hand side is smaller than the right-hand side, their quotient being $1/\sqrt{\pi n}$. In fact, even the relative entropy cannot strictly satisfy the principle that the entropy of the whole is the sum of the entropies of its parts, for the reason that the probability of the whole is not strictly equal to the product of the probabilities of its parts. This decrease of the probability with the opening of the window is explained physically by the fact that now the number of molecules in each of the two containers can change; the molecules may accumulate in one container. But this possibility need not be taken into consideration, and the neglect of the third term in Stirling's formula appears therefore justifiable.[4]

_____

[4] It might be suggested that we regard the neglect of the last term in Stirling's formula not as an approximation, but as a matter of definition for the mathematical form of entropy. Then the relative entropy would strictly satisfy the additivity postulate. However, this way out appears inappropriate, because we then cannot say that the probabilities of actual occurrences are strictly governed by the entropy concept.

● 9. Extension of Statistics to Different Energy Levels

We shall now turn to extending the entropy definition to gas systems in which the cells, or compartments, are associated with different energy levels, as illustrated by the various values of kinetic energy which a molecule can have as a function of its velocity. Since molecular speed is related to temperature, the study of the statistics of molecular velocities and the definition of a corresponding entropy allows us to describe the process of temperature equalization in a gas in which, at the beginning, the temperature is unevenly distributed. We will follow Boltzmann's presentation.

Regarding the three components v_1, v_2, and v_3 of the velocity of a molecule as Cartesian coördinates, Boltzmann constructs a *velocity space*, which he divides into a large number m of cubical cells of equal size, corresponding to those used in the previous example. Each cell can be conceived as a little cubicle $d\omega = dv_1 dv_2 dv_3$. A molecule which at a certain moment t has a certain velocity, is regarded as occupying one of the cells of this fictitious space. A cell can contain many molecules, all those, namely, which at the moment t have the same velocity within the interval of exactness defined by the cells. If for each cell the number n_i of occupying molecules is given, we know the *velocity distribution* of the molecules at the time t. At another time t' there may exist a different velocity distribution of molecules.

Every velocity distribution, and every arrangement, must satisfy condition (8, 1), which we shall repeat here:

$$\sum_{i=1}^{m} n_i = n \ . \tag{1}$$

But, in addition, it must satisfy the condition that the kinetic energies of the individual molecules add up to the heat content of the gas, since heat energy is nothing but the sum total of the mechanical energies of the individual molecules. The kinetic energy ϵ of a molecule of mass μ is given by

$$\epsilon = \frac{\mu}{2} \left(v_1^2 + v_2^2 + v_3^2 \right) \ . \tag{2}$$

This quantity varies continuously with v_1, v_2, v_3; but since we assumed that the ranges of the latter variables are divided by steps so as to form little cubicles, we correspondingly divide the range of ϵ into steps $d\epsilon$. We thus construct energy levels $\epsilon =$ constant, which are defined to the exactness $d\epsilon$. Each cell belongs to some energy level; and each energy level ϵ includes a number of cells. Calling ϵ_i the energy level associated with the cell i (note that different cells i and j may have the same energy level, such that $\epsilon_i = \epsilon_j$), we can write the energy condition in the form

$$\sum_{i=1}^{m} n_i \, \epsilon_i = E \ . \tag{3}$$

E is the total energy content of the gas.

If an arrangement satisfies conditions (1) and (3) it may be called a *possible arrangement.* The number N_E of possible arrangements is finite, because the energy condition limits the velocities of the molecules. A molecule possesses the maximum velocity v_{max} if it carries all the heat energy of the gas, while every other molecule has the speed zero.

As in the preceding example, a probability metric must now be introduced. It can be described as follows. The highest speed v_{max} limits the cells in the velocity space to the interior of a sphere K of the radius v_{max} around the origin of the coördinates. Let us say that the sphere includes m cells. As before, there thus are $N = m^n$ arrangements. Most of these will contradict the energy condition formulated in (3); therefore, all such arrangements are canceled. The remaining N_E arrangements, which satisfy the energy condition, are now regarded as equally probable and are assigned the probabilities $1/N_E$. This is the probability assumption on which Boltzmann's theory is based; we see that this theory presupposes a constant probability metric in the velocity space, subject to the restriction by the energy condition.

The number N_D of arrangements within a distribution is again given by (8, 5). We thus have here

$$W = \frac{N_D}{N_E} \ . \tag{4}$$

The number N_E does not admit of a simple computation. Using (8, 7), we thus write the entropy, in analogy to (8, 8), as follows:

$$S = \log W = n \, \log n - \sum_{i=1}^{m} n_i \, \log n_i - \log N_E \ . \tag{5}$$

Since we are interested essentially only in the maximum of W, that is, in the distribution which, for a given E, corresponds to the state of equilibrium, it does not matter that the constant N_E in (5) is unknown, and we can treat (5) as an absolute entropy, disregarding the constant. The maximum of S in (5) is found, as before, by maximizing the summation term, this time, however, by the use of the two restrictive conditions (1) and (3). Lagrange's method leads to the result that

$$\sum_{i=1}^{m} \left[\delta\left(n_i \log n_i\right) + \delta\left(\alpha n_i\right) + \delta\left(\beta \epsilon_i n_i\right) \right] = 0 \ ,$$

$$\log n_i + 1 + \alpha + \beta \epsilon_i = 0 \ , \tag{6}$$

$$n_i = a e^{-\beta \epsilon_i} \ , \qquad a = e^{-1-\alpha}.$$

Since α and β are constants, this formula shows that the occupation number n_i of a cell depends only on the energy level associated with the cell. In application to the velocity space, this means that n_i depends on the kinetic energy corresponding to the values v_1, v_2, v_3 belonging to the cell. Let $m(\epsilon)$ be the number of cells associated with the same energy level ϵ, and $n(\epsilon)$ the total number of particles in these cells when the gas is in equilibrium; then we have, because n_i is constant for these cells,

$$n_i = \frac{n(\epsilon)}{m(\epsilon)} = q(\epsilon) = a e^{-\beta \epsilon} \ . \tag{7}$$

Here $m(\epsilon)$ is a constant for the gas system. The quantity $q(\epsilon)$ is the number of particles per cell, and thus the occupation number for any cell, for the energy level ϵ in the state of equilibrium. The subscript "i" on "ϵ" has been omitted, because ϵ no longer refers to a particular cell, but to an energy level.

Postponing the discussion of the constant a, we will now evaluate the constant β, that is, express it in thermodynamical parameters. First, we can give (8, 9) a new form by replacing "log n_i" by "log $a - \beta\epsilon_i$", a value which follows from (6). We find, using (1) and (3),

$$S = k \left[n \log n - n \log a + \beta E \right] \ . \tag{8}$$

Second, we can now combine (8), which expresses the dependence of S on the total energy E, with the fundamental relation $(7, 1)$ of non-statistical thermodynamics, in which the amount of heat Q represents the energy E. Differentiating (8) and then using $(7, 1)$, we find

$$\frac{dS}{dE} = k\beta = \frac{1}{T} \ ; \tag{9}$$

therefore,

$$\beta = \frac{1}{kT} \ . \tag{10}$$

Inserting this value in (7), we arrive at the result

$$n_i = \frac{n(\epsilon)}{m(\epsilon)} = q(\epsilon) = a e^{-\epsilon/kT} \ . \tag{11}$$

The quantity $q(\epsilon)$, which, for the state of equilibrium, measures the number of particles per cell, is thus shown to depend in the exponential form (11) on the energy level ϵ and the temperature T. Since T is a property of the gas system as a whole, relation (11) says that differences between the occupation numbers of different energy levels depend on the condition of the gas as a whole. If the temperature T of the gas is raised, the occupation number of each cell in a specified energy level becomes larger in relation to the occupation numbers of cells at lower energy levels; the molecules leave the levels of lower energy and move into those of higher energy.

Although we derived (11) for a gas system in which the different energy levels originate merely from differences in the speed of the molecules, the result (11) is not dependent on this special case. In fact, (11) is derivable whenever (8,5) represents a measure of the probability of a statistical state and the system is closed, an instance in which the energy condition assumes the form (3). And that the relation (8,5) measures the probability, follows from the two presuppositions of the equiprobability of arrangements and the individual distinguishability of the elementary particles. Statistics satisfying these two presuppositions are usually called *Maxwell-Boltzmann statistics*. For this kind of statistics, the exponential dependence of the stationary occupation number $q(\epsilon)$ on energy level and temperature expresses a fundamental law, which applies whether the energy levels result from molecular speed alone or from speed in combination with a field of gravity (as existing for the earth's atmosphere), or from other sources.

We have regarded the argument "ϵ" in (11) as a continuous variable, whereas the quantity n_i refers to cells $d\omega$ of a certain size. In order to eliminate this inconsistency, we introduce the *continuity assumption* that, going to smaller $d\omega$, the following limits exist:

$$\frac{n_i}{d\omega} \to n_i' \ , \ \frac{n(\epsilon)}{d\omega} \to n'(\epsilon) \ , \ \frac{q(\epsilon)}{d\omega} \to q'(\epsilon) \ , \ \frac{a}{d\omega} \to a' \ . \tag{12}$$

The quantity $m(\epsilon)$, in contrast, remains constant when $d\omega$ goes to 0. Postulating the limits (12) involves some difficulty, because, going to smaller $d\omega$, the quantities n_i become either 0 or 1. This difficulty can be overcome when we interpret n_i as the time average of the occupation number. Or we may say, as is common in physics, that we do not mean here a strict limit resulting when $d\omega$ goes strictly to 0, but use a practical limit which manifests itself when variations of $d\omega$ within a certain range of smallness are considered. Associating the transition to smaller $d\omega$ with an increase in the particle number n is permissible only when a simultaneous increase in volume is assumed, and thus could be referred to for the velocity space, but not for physical space. Otherwise, the increasing gas pressure would lead to a change in physical conditions.[1]

When we now divide n_i, $n(\epsilon)$, $q(\epsilon)$, and a, in (11), by $d\omega$, we find the same relation as holds for the primed quantities. Letting

$$q'(\epsilon) = n \cdot p(\epsilon) , \qquad a' = n \cdot b , \qquad (13)$$

and introducing the form (2) for ϵ, we find, now regarding p as a function of v_1, v_2, v_3:

$$p(v_1, v_2, v_3) = b \cdot e^{-(\mu/2kT) \cdot (v_1^2 + v_2^2 + v_3^2)} . \qquad (14)$$

This relation, which formulates *Maxwell's velocity distribution*, states that the probability $p\,d\omega$ of finding a molecule in the cell $d\omega$, for the state of equilibrium, is given by a normal curve, or rather, a normal surface, in three dimensions, namely, in the three coördinates v_1, v_2, v_3. The constant b can now be evaluated by the condition that integrating (14) over dv_1, dv_2, dv_3 must yield the value 1. The integration furnishes the result[2]

$$b = \left(\frac{\mu}{2\pi kT}\right)^{3/2} . \qquad (15)$$

In contrast to the constant β of (8), which allows a general evaluation, we see that the constant $b = n \cdot a'$ can be ascertained only after the functional form of the energy ϵ in terms of certain parameters is given, as in (2). The bell-shaped surface has its vertex at $v_1 = v_2 = v_3 = 0$;

---

[1]Using the limits (12), we can define the value of the absolute entropy in a form which does not depend on the size of the cells $d\omega$. For the relative entropy, this condition is automatically satisfied. When we compute the entropy for smaller cells, applying the continuity assumption mentioned above, the relative entropy remains the same although the number of cells increases: for instance, for cells of half size, their number increases to $2m$, but we find that the relative entropy (8, 8) does not change. However, if we should omit the third term in (8, 8), the entropy would change. This can be remedied by adding a term $n \log d\omega$ to the constant and including it in the summation term. The absolute entropy can then be written in the form

$$S = n \log n - \sum_{i=1}^{m} [n_i' \log n_i'] \, d\omega + \text{const} ,$$

in which equation n_i' is the limiting value defined in (12). This form makes it possible to replace the summation by an integration.

[2]See, for instance, Richard C. Tolman, *The Principles of Statistical Mechanics* (Oxford, The Clarendon Press, 1938), p. 90.

this means that all directions of the velocity are equally probable, and thus that their mean value equals 0. It is different for the mean value of the speed v, that is, the absolute amount of the velocity. Relation (14) furnishes for the root-mean-square v_0 of the velocity the value[3]

$$v_0 = \sqrt{\overline{v^2}} = \sqrt{\frac{3kT}{\mu}}. \tag{16}$$

This result shows that the most probable distribution is not represented by equal numbers n_i in all the cells of the velocity space, but by a distribution in which many molecules have an average speed v_0, while some have a larger and some have a smaller speed. The average speed v_0 increases with the square root of the temperature of the gas as a whole, according to (16). Larger deviations from the average value, on both sides, are less frequent than small ones. But, as before, this distribution in velocity space is characterized by having the largest number of arrangements; and equality of probabilities for all possible arrangements is thus shown to lead to unequal probabilities for the various distributions and to the distinction of one as the most probable distribution.

From the computation of probabilities we again proceed to a conclusion concerning temporal changes: if the molecules are originally arranged in a distribution of low probability, and if they are then stirred or agitated, they will proceed to distributions having higher and higher probabilities. The shuffling mechanism is provided by the collisions of the molecules; the study of this effect of collisions on the exchange of speeds is one of Boltzmann's chief contributions. If in the beginning the gas is separated into a cold part and a warm part, the collisions of molecules tend to produce an equalization of molecular speeds; and eventually the gas attains a state of thermodynamical equilibrium, that is, a state of maximum probability in which the velocities of the molecules are distributed according to a normal curve, as formulated by the distribution (14). Once this distribution has been reached, it remains stationary; that is, though in further collisions the velocities of the individual molecules are exchanged, the distribution as a whole remains unchanged. Thus the probability interpretation supplies again the explanation of the evolution toward equilibrium—in this example, toward temperature compensation.

We have developed the entropy concept in two applications: first, in reference to cells in the physical space occupied by a gas; and second,

---

[3]For the computation, see *ibid.*, p. 92.

in reference to cells in the velocity space. It is easy to combine the two forms of statistics into one. A combination of the three dimensions of physical space and the three dimensions of the velocity space will form a six-dimensional parameter space. When we say that a molecule is in a certain cell of this space, we mean that it is at a certain place of physical space and that its velocity has a certain value. The definition of entropy is exactly the same as before. In fact, we can take over the forms (8, 8), (8, 9), and (8, 11) without change. The summation extends then to the number m of cells in the parameter space used. The maximization method, carried out above in (8, 13) and (6) separately, would then be applied in one procedure to the entropy defined for the parameter space, and would yield the same results. This consideration shows that entropy can be defined as a comprehensive function which characterizes the state of the gas as a whole.

Some remarks concerning the term "state" may now be added. We distinguish between *microstate* and *macrostate*. A microstate is the same as an arrangement. A macrostate is a joint class, or disjunct, of distributions resulting when slight variations of the numbers n_i of molecules in the cells are admitted, if only these variations remain unnoticeable for macro-observations. A macrostate A is therefore a class of microstates a which are *macroscopically indistinguishable* from one another. For instance, if two gases are well mixed, in such a way that equal small subvolumes contain approximately the same number of molecules from each gas, this macrostate A can be realized by various microstates a; in fact, this is a macrostate of equilibrium which remains unchanged while its microstates change continually. If A is not a state of equilibrium, it will also be realized by many states a. For instance, if we open the cock of a gas stove for a moment, the gas will at first be found only in the environment of the stove. This is a macrostate A which can very well be macroscopically distinguished from a state of complete mixture of gas and air in the room. But, again, A can be realized by various microstates a.

The distinction between microstates and macrostates, which has always been clearly carried through in the literature on statistical mechanics, leads to a natural grouping of microstates. Given a microstate a_1, the macrostate A to which it belongs is well determined: A includes all those microstates a_i which are macroscopically indistinguishable from a_1. We call A the macrostate *corresponding* to the microstate a_1, or, briefly, the *corresponding macrostate*. This classification allows us to speak occasionally of the entropy of a microstate a; we then refer not to the probability of a, but to the probability of the corresponding macrostate A. If this interpretation is used, the phrase *entropy of a microstate* can be employed without danger. In the same sense we shall also speak of the macroprobability of a state *a*, meaning by this phrase the probability of the corresponding macrostate *A*.

A well-shuffled state is sometimes called an *unordered* state, in contrast to an *ordered* state, such as is exemplified when all the nitrogen molecules are on one side of a container and all the oxygen molecules on the other side. The term "unordered" used in this sense does not always correspond to the usage of everyday life. With reference to the illustration in which the gas is found mainly in the environment of a stove soon after the cock was opened for a moment, one might argue that we would call this condition of air and gas a rather disorderly state, compared with which an equal distribution of the gas over the room appears highly ordered. However, this is not the meaning of the term "order" of statistics. By "order" we understand here the *degree of separation;* therefore, a well-shuffled distribution possesses the lowest degree of order, whereas a partial separation has some higher degree of order. The term "separation" may refer to a separation of molecules of different substances, or of different velocities, or of other properties which, through separation, become macroscopically observable.

Using this meaning of "order", we may say that macroscopically indistinguishable arrangements possess *the same degree of order*. What we called above the corresponding macrostate of an arrangement may therefore also be called its *same-order class*. The degree of disorder then can be measured by the number of arrangements in this class; that is, if there exist a large number of same-order arrangements for a given arrangement *a*, the order of *a* has a low degree. Conversely, a high degree of order means that the same-order class is small. The concept *order* is therefore definable in terms of the concept *same order*. which in turn is reducible to the concept *macroscopically indistinguishable;* and the degree of order is given by the extension of the same-order class. This is the conception on which the interpretation of entropy as a measure of disorder, or an inverse measure of order, is based, since entropy, and probability, increase with the number of arrangements belonging to a state.

● *10. A Deterministic Interpretation of*
    Thermodynamical Statistics

The considerations given in the preceding section show that probabilities are determined, in the theory of gases, by an enumeration of the arrangements in the same-order class. It was pointed out that this method presupposes the equiprobability of all arrangements, as explained above. The probability computations of statistical mechanics, therefore, are based on an assumption concerning an *initial probability metric*. The empirical nature of this assumption is clear from the fact that we regard changes in the state of a gas as being controlled by this probability, that we expect the history of a gas to reflect the assumed probability metric: most of the time a gas is found in probable states, and seldom in improbable ones. Only because of this translation into lengths of time intervals can we claim to have found in the probabilities of macrostates the determining factor which accounts for the transition from high-order to low-order states.

The use of an initial probability metric is indispensable if one desires to derive statistical conclusions. Yet the question how to account for this metric has always bothered the statisticians. Rationalistically inclined mathematicians, like Laplace, have believed that the initial metric can be justified on a priori grounds. Although some contemporary logicians still adhere to similar conceptions, the majority of statisticians and mathematical physicists have rejected any a priori introduction of equal probabilities. If cells of equal size in the velocity space are assumed to be equally probable, physicists would argue that such an assumption must not be based on any kind of logical argument, as, for instance, the principle of indifference, but can only be validated by an experimental test of its consequences. And they may refer to the fact that in the domain of very low temperatures this metric has turned out to be invalid and has been replaced by another one. That is, a probability metric can only be introduced in the sense of a physical hypothesis and has to wait for confirmation by observations. But, it should be noted, the logical theory of confirmation must not then, in turn, itself be based on a metrical probability assumption —a conclusion which certain modern logicians have forgotten to draw.

Yet there remain other than aprioristic ways for the physicist to account for a probability metric even before specific tests have been made. He may attempt to derive this probability metric from the causal laws of physics. This would be a legitimate method, because it merely transfers the results of one domain of experience to another. Such derivations can be made in two forms. In the first form, they start from a set of assumptions which include, in addition to the causal laws, certain general probability assumptions that do not presuppose a probability metric. Of this form, for instance, is a derivation constructed by Henri Poincaré for the equiprobability holding in games of chance.[1] In the second form, the derivation starts from causal laws

[1] See *ThP*, §69.

only and does not include any probability assumptions. The derivation constructed for the probability metric of the kinetic theory was originally of the first form. But it has been the ambition of mathematical physicists to eliminate probability assumptions completely. How near they have come to this goal will be seen presently. We shall now study these derivations in more detail, beginning with the first form.

Investigations of this kind were undertaken by Ludwig Boltzmann, J. C. Maxwell, J. W. Gibbs, and others. These mathematical inquiries are usually attached to a more general form of describing molecular states. A molecule is regarded as a system of m position parameters $q_1 \ldots q_m$ and, likewise, m momentum parameters $p_1 \ldots p_m$. These parameters include not only velocity and spatial location, but possibly other quantities, such as quantities referring to the rotation of the molecule. For the whole gas, the total number of parameters is $N = 2mn$, where n is the total number of molecules in the gas, and this number may be counted through as $x_1 \ldots x_N$. The instantaneous state of the gas, its microstate, can be pictured geometrically as a point in the N-dimensional *phase space*, built up by the generalized coördinates $x_1 \ldots x_N$. Since the microstates of the molecules change from moment to moment, the phase point travels along in the phase space. Let us assume that the gas system is *closed*, that is, completely cut off from any interaction with its environment. Then the path of the phase point, that is, the passage of the gas system from microstate to microstate, is strictly determined by the laws of mechanics. This means: given a point in the phase space, there exists precisely one path starting from it; this path goes on and on, winding its way through the space.

The causal laws determining the path are the equations of motion; they are usually written in the *canonical* form, introduced by W. R. Hamilton:

$$\frac{\partial H}{\partial p_i} = \frac{dq_i}{dt} , \qquad \frac{\partial H}{\partial q_i} = -\frac{dp_i}{dt} . \qquad (1)$$

These partial differential equations, in which H, the Hamiltonian, represents the total energy of the system, are generalized forms of Newton's fundamental laws of mechanics.

One of the consequences of equations (1) is the conservation of energy. Since this condition is a relation of the form

$$E\left(x_1 \ldots x_N\right) = \text{const} , \qquad (2)$$

it defines an $N-1$-dimensional hypersurface in the phase space, the energy surface, to which the travels of the phase point are bound.

The method adopted by Boltzmann and others can now be characterized as follows. Although the path starting with a given point is
strictly determined by equations (1) for any length of time, it is regarded
as a sequence of states having the properties of a probability sequence.
In its course, it passes through all possible states with equal probabilities, and it stays most of the time in unordered states. We thus
have here a combination of a deterministic and a probability interpretation of physical occurrences: the path of the phase point is at one
and the same time a strictly predictable causal chain and yet a probability sequence. The phase point travels in a zigzag path along the
energy surface like a toy automobile which, once wound, chases along
determined curves on the floor; and like this little mechanical contriv-
ance, it owes its varying positions to hidden laws driving it inescap-
ably from place to place.

Let us study in more detail this probability sequence run through by
the phase point. We have here a continuous probability sequence. Such
sequences have an aftereffect;[2] that is, when the phase point is at the
place x, it is to be found a moment later still in the environment of x.
But gradually the point travels into more distant parts of the energy
surface; and if we do not know its past, we can expect to find the point
at any place on the surface. There exists, therefore, for such a sequence, an *over-all probability* which has a well-defined value for any
area of the energy surface; that is, given an area A of the surface, we
can speak of the probability of finding the phase point within A irre-
spective of its previous positions.

Using terminology from the calculus of probability, we call the
energy surface the *attribute space* of the sequence. If we describe this
surface by means of orthogonal coördinates $y_1 \ldots y_{N-1}$, which are
located within the surface, we have in the attribute space a distribution function

$$\varphi\left(y_1 \ldots y_{N-1}\right) \tag{3}$$

the integral of which over a certain area A measures the over-all probability of finding the phase point in this area. If A is a macrostate as
defined above, we thus find its entropy by integration of (3). What we
have called a microstate a, or arrangement, is represented by a very
small area $dy = dy_1 \ldots dy_{N-1}$; its probability results from (3) simply
by multiplication with dy, and is thus very small. Its same-order class,
however, may occupy a large area. That unordered states are highly
probable means that they cover a very large area of the energy surface.

---

[2] See §12 of the present study.

Probability sequences that are given by a rule determining the attribute for every element offer no difficulties to the theory of probability.[3] Examples of discontinuous sequences of this kind have repeatedly been studied. With respect to a table giving the square roots of positive integers to the 5-th decimal place, for instance, it has been shown[4] that in the last digit, odd and even numbers are equally probable, or, in other words, that the over-all probability for even numbers equals 1/2. However, the series of 5-th digits has an aftereffect, that is, for larger integers we get longer and longer runs of even numbers, and of odd numbers, because the initial sections of the decimals then change rather slowly from integer to integer.

If a probability sequence is given by a rule, it is of course to be required that its probability distribution be derived from the defining rule, as has been done for the example just given. Physicists have therefore made many efforts to construct a proof that the equations (1) of motion lead to the probability metric assumed in the kinetic theory.

The nucleus of all these proofs is a theorem named in honor of the French mathematician Liouville, who in 1838 investigated the condition of incompressibility for a liquid. Assume that we have a domain A in the phase space; from every point of A we construct the path of the phase point for a time interval Δt, which interval is the same for all points. The ends of the paths are then found within a domain A' (fig. 13). We can say: A is mapped onto A' by the phase path in the time Δt. Since there is only one path starting from a given point in A, the mapping is completely determined by the mathematical relations laid down in equations (1). We can also give the mapping a physical illustration: We regard each point in A as represented by a separate closed system at the time t; after each system has gone through its series of varying molecular arrangements up to the time $t + \Delta t$, the totality of the systems fills the domain A'. By the use of equations (1), it can now be shown that the domains A and A' have equal volumes.[5] Comparing the motion of all these phase points to a current in a liquid,

---

[3] *ThP*, pp. 132, 165, 343.

[4] See György Pólya and Gábor Szegö, *Aufgaben und Lehrsätze aus der Analysis*, Bd. I (Berlin, J. Springer, 1925), pp. 72 and 238. Henri Poincaré has made a corresponding remark for a table of logarithms, but it has been shown by J. Franel that Poincaré's proof is incorrect (see Pólya and Szegö, *op. cit.*, pp. 73 and 239). Not knowing of this criticism at the time, I quoted Poincaré's example in *ThP*, p. 343; this example should be replaced by the one above concerning square roots.

[5] For the proof, see any textbook on statistical mechanics; for example, Richard C. Tolman, *The Principles of Statistical Mechanics* (Oxford, The Clarendon Press, 1938), p. 49.

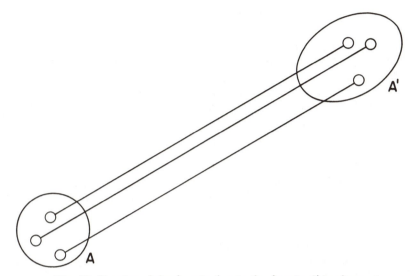

Fig. 13. Mapping of the domain A onto the domain A' in phase
space, performed by the paths of phase points.

we can formulate Liouville's theorem as stating that this liquid is
incompressible, and that its parts maintain their volumes during their
flow. In the physical illustration given, this means that the course of
changes in the assemblage of separate closed systems resembles, in
its mathematical relationships, a current in an incompressible liquid.

This theorem leads to an important conclusion concerning the path
of only one phase point. Assume that the path enters the domain A
and stays in it for a time period t_A. Now we know that when the point
is in A, it will be in A' after the time Δt; it will then stay in A' for
exactly the same time as in A. That is, we have $t_A = t_{A'}$. If the phase
point enters and leaves A repeatedly, the same will hold for the domain A'; every stay in A is followed by a stay in A' after the time Δt.
The sum of all time lengths t_A will therefore equal the sum of the time
lengths $t_{A'}$. Since, according to Liouville, A and A' have equal volumes,
we conclude that the phase point stays in equal volumes for equal
times.

This equality of *times of sojourn* supplies the basis for the probability metric assumed in gas theory. For instance, it leads to the
equiprobability of cells in the velocity space, which, in turn, leads to
low probabilities for ordered states; such states occupy only a small
volume in the phase space. The equiprobability condition assumed by
Boltzmann is thus shown to be a consequence of the equations of
motion.

However, the result explained cannot be used without some important additions. We have to assume, first, that the phase point enters eventually every volume element, however small we choose it. Otherwise, there might be volume elements equal in size to A which would never be reached by ergodic paths from A, and for which, therefore, the equality of times of sojourn could not be asserted. Furthermore, we must assume that the total sojourn time Σt_A in A maintains a certain ratio to t, that is, that

$$\lim_{t \to \infty} \frac{\Sigma t_A}{t} = \text{const} . \tag{4}$$

This assumption allows us to speak of the probability that the phase point is to be found within the domain A; it defines this probability, by analogy with the frequency interpretation of probability, as the limit of a ratio of time periods.[6] The constant in (4) has to be the same for every path that passes through A. Such an assumption is necessary, because the individual pieces of path traversing A may belong to different total paths.

An assumption of this kind was introduced by Boltzmann under the name of *ergodic hypothesis* (*ergos* = energy, *odos* = path). Boltzmann formulated it as the hypothesis that the phase point passes eventually through every point of the energy surface. This formulation is easily shown to be untenable. It was replaced by P. and T. Ehrenfest by the formulation that the path comes close to every point within any small distance ϵ which we select and which is greater than 0.

There still remained the question whether the ergodic hypothesis must be regarded as an independent presupposition, or whether it is derivable from the canonical equations (1), as Liouville's theorem is. If the first alternative were true, we would have here a derivation of the first form, as mentioned above, in which a probability assumption, the ergodic hypothesis, is added to causal laws, namely, Liouville's theorem, in order to achieve the derivation of the probability metric. If the second alternative were true, we would have here a derivation of a probability metric from causal laws alone.

This problem has occupied mathematicians for a long time. It was finally solved through ingenious investigations by John von Neumann and George Birkhoff,[7] who were able to prove that the second alternative

---

[6] For this definition of probability for continuous sequences, see *ThP*, p. 240.

[7] See, especially, six articles published in the *Proceedings of the National Academy of Sciences*: In Vol. 17 (1931)—George D. Birkhoff, "Proof of the Ergodic Theorem", pp. 656–660. In Vol. 18 (1932)—John von Neumann, "Proof

is true. The idea of this derivation of the ergodic hypothesis from the equations of motion can be illustrated by the following consideration. Assume that a billiard ball is pushed from A (fig. 14) in the direction toward B, and assume that there is no friction, and hence that the ball, reflected by the wall at B, continues its path without stopping. Because of further reflections by the walls, the ball will travel a zigzag path which eventually will cover the whole surface of the billiard table. This would be a special and very simple illustration of Von Neumann and Birkhoff's theorem, for which the phase point corresponds to the ball and the energy surface to the billiard table.

Only for exceptional initial conditions can it happen that the ball's path is restricted to a part, or a subsurface, of the total surface (fig. 15). However, Von Neumann has shown that then his theorem applies to this subsurface taken separately, which, again, will be covered with equal density by the phase point.[8]

With Von Neumann and Birkhoff's theorem, deterministic physics has reached its highest degree of perfection: the strict determinism of elementary processes is shown to lead to statistical laws for macroscopic occurrences, the probability metric of which is the reflection of the causal laws governing the path of the elementary particle. Unfortunately, this triumph of classical mechanics was achieved at a time when it had already been replaced by quantum mechanics, which no longer recognizes the determinism of elementary processes. But let us postpone the discussion of quantum physics and first go on studying

---

of the Quasi-ergodic Hypothesis", pp. 70–82, and "Physical Application of the Ergodic Hypothesis", pp. 263–266; Eberhard Hopf, "On the Time Average Theorem in Dynamics", pp. 93–100, and "Complete Transitivity and the Ergodic Principle", pp. 204–209; and George D. Birkhoff and B. O. Koopman, "Recent Contributions to the Ergodic Theory", pp. 279–282.

[8]Sometimes, assumption (4) is called *ergodic theorem* and thus distinguished from the *ergodic hypothesis*, according to which the phase point enters every volume element. Whether the ergodic hypothesis holds for a given total space depends on the initial conditions. The proof given by Von Neumann then can be formulated as follows: There is always some subspace for which the ergodic hypothesis is true; and for this subspace, the ergodic theorem is always satisfied. Since the ergodic theorem is trivially satisfied for volume elements into which the phase point does not enter, the constant in (4) then being equal to 0, this theorem can be asserted unconditionally. Incidentally, though the probability metric in the phase space is constant, it is given for the energy surface by another function, the ergodic density σ. The reason is that neighboring energy surfaces do not have a constant distance from one another. It can be shown that σ is the reciprocal value of the gradient of energy.

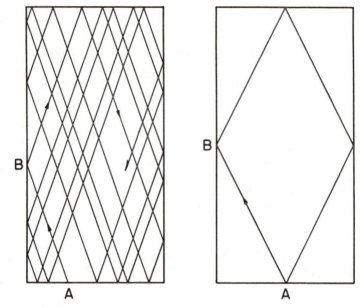

Fig. 14 (left). Path of a billiard ball reflected at walls and covering eventually the whole table.

Fig. 15 (right). Exceptional path of a billiard ball, forming a closed cycle.

the classical solution of gas-statistical problems. We shall see that a clear conception of the classical form of gas statistics is indispensable for understanding the contributions of quantum mechanics toward a solution of the problem of time.

Even in classical physics, however, the deterministic solution of gas statistics can be questioned. What has been proved is that the statistical laws governing the macrocosm are *compatible* with strict laws governing atomic occurrences; but this does not mean that the deterministic assumption supplies the *only* solution to the problems of gas statistics. We can very well assume that atomic occurrences are not strictly predictable in terms of causal laws, whereas macrocosmic occurrences, in which large numbers of molecules are involved, are controlled by statistical laws of a high degree of precision. We can carry through a conception of this kind without contradicting the laws of classical physics; in fact, we do apply such conceptions whenever we are concerned with individual phenomena of thermodynamics.

Consider, for instance, the application of the ergodic theorem to some closed physical system used in the laboratory. We know that perfect thermal insulation cannot be achieved. The phase point of the system will therefore not be determined by the equations (1) of motion alone; it will be subjected repeatedly to irregular little impacts from the outside. These chance impacts, though of small effect in the beginning, will change the later path of the point considerably; in particular, they will push the point out of a restricted subsurface whenever it so happens that the phase point occupies an exceptional position p_0. For such reasons, for instance, it will never be possible to have the process diagramed in figure 15 go on for a very long time; even if we could arrange the initial conditions correctly, disturbances from the outside would make the ball leave its exceptional path, which would then assume the general form presented in figure 14.

It is therefore a result of the chance impacts from the outside that the sequence run through by the phase point is not strictly determined; yet the sequence can still be a probability sequence with aftereffect, satisfying all the conditions required for the results of gas theory. The little disturbances by the environment, subject only to the laws of chance, represent a sort of monkey wrench thrown into the gears of Hamilton's equations; but they do not disturb the statistical results derived from these equations, because a probability sequence, or probability chain, can possess all desirable statistical properties without having any deterministic linkage between its successive elements. All probability sequences presented by individual physical systems are of this indeterministic kind, because there exist no completely isolated systems.

It may be argued that if there are impacts from the outside, we may include them in the system considered by extending the outer boundaries of the system. The disturbances thus are made changes of parameters controlled by the laws of motion and, therefore, by the rules governing the ergodic path. The only difference will be that with the new parameters some further dimensions have been added to the phase space; but the path of the system will again be strictly determined. If there are further disturbances from the outside, this extension may be continued. Finally we shall thus come to a system which is identical with the universe. The most comprehensive system is completely isolated and should therefore be controlled by strict causal laws, undisturbed by impacts from the outside.

This is the hypothesis made by determinism. We must now inquire what reasons can be adduced for and against determinism.

● 11. Determinism Versus Indeterminism in Classical Physics

Determinism is not an observational fact. It is a theory derived from observations by way of an extrapolation. Since observations lead to physical laws of great predictive power, it is assumed that if we could make more precise observations we could make predictions that would come true without exception. A set of ultimate causal connections is supposed to be hidden behind observable relationships. Determinism is thus based on an extension of observed regularities to unobserved ones; and it is assumed that the flaws of attainable predictions would vanish if we could only uncover the ultimate causal structures.

Let us analyze this inference in more detail. Suppose we wish to predict the place where a shell fired by a gun will hit the ground. We measure the direction of the gun's barrel, the amount of the powder charge, the mass of the shell, and so on; and using these values as given *parameters*, or *initial conditions*, of the process, we compute the point of impact by the use of physical laws. It is well known that the prediction thus made is none too reliable. Why is it not reliable?

Because, we would argue, we cannot include all the relevant factors in our computations. We can only say: If the parameters referred to have the assumed values, and if no other parameters have any influence on the process, the shell will arrive at the predicted place. That is, we use the assumption of causality in a *conditional* form. And we know very well that this form is not sufficient to guarantee the truth of the prediction. The if-clauses will not be completely satisfied; the measured values of the parameters will hold only to some degree of exactness, and there will be further parameters, neglected in the computation, which may have an influence on the occurrence. Some of these might be included in a more precise calculation; for instance, we might take into consideration the rotation of the earth and the wind conditions. Others we cannot take into account, because human abilities of observation are limited. For instance, a meteor might arrive from the depths of the universe and hit the shell on its path; then our prediction would be completely invalidated.

It would not help us to qualify our prediction by the condition: "if no disturbances intervene". Predictions thus qualified have no practical use, unless we know that disturbances and other deviations are exceptional occurrences. We want to be sure that most of the predictions based on the parameters considered are true, that is, that any such prediction is highly probable. We therefore must supplement the conditional form of causality by a probability hypothesis, according to which we can select the parameters to be considered in such a way that they will permit a prediction of high probability even though the if-clauses of the conditional form are not strictly satisfied. Only in its combination with the probability hypothesis is the conditional form of causality a nonempty statement which is testable through observations. In this sense, the probability concept is indispensable even for classical physics.

We know from many experiences that this probability hypothesis is true; and we also know that we can improve the probability of the prediction by including more and more parameters in the computation. Often it is necessary, furthermore, to replace the physical laws used by more precise ones. For instance, the laws controlling the friction of the shell in the air are only of an approximate nature and are subject to improvement. In view of such considerations, we have come to believe that causality can be extended to an *unconditional form:* We assume that it is physically possible, though not technically possible, to know all the parameters of the process and to know the ultimate physical laws. We then conclude, using the conditional form, that on this assumption we could make a prediction with the probability 1, that is, with practical certainty. But it is the unconditional form of causality which expresses *determinism* and which we must analyze. We shall thereby proceed on the assumption that the conditional form, at least, is true, and thus remain within the framework of classical physics. In quantum physics, even the conditional form has been abandoned; but we shall postpone the discussion of this turn of the problem to chapter v.

Because of its reference to ultimate initial conditions and ultimate laws, determinism is a limit statement. As such it has meaning only if it is translatable into statements of convergence referring to actual observables and to actually known laws. It was pointed out that these convergence statements involve the concept of probability; they must now be formulated precisely.

If $D^{(1)}$ describes the physical state of a spatial volume v at the time t_1, we can predict with a certain probability $p^{(1)}$ that the volume v will be in a certain state described by E at a later time t_2. In order to raise this probability to a higher value $p^{(2)}$, we replace $D^{(1)}$ by a more precise description $D^{(2)}$, which likewise refers to the time t_1. This new description can differ from $D^{(1)}$ in any or all of the following respects:

1. The new description includes the initial environmental conditions of the volume v; that is, it refers to an extended spatial volume V. However, the description E still refers to the original volume v.

2. The new description refers to more precise measurements of the parameters and employs inner parameters of the system included in v which had been neglected—for instance, parameters referring to the inner state of the molecules.

3. The new description uses improved causal laws.

This procedure may be continued. The descriptions $D^{(i)}$ will then be changed from macrostate descriptions to microstate descriptions. In classical physics, it was usually assumed that with respect to points 2 and 3 we soon reach a final stage, which leads to ultimate precision, and that only point 1 involves difficulties of a fundamental nature. The ultimate laws were believed to have been found in the equations of motion; and the ultimate parameters, in a mechanical model of the atom. It is an interesting fact that as early as 1898 Boltzmann had questioned this belief with respect to points 2 and 3.[1] Today one would scarcely be willing to grant the assumption that the ultimate laws have been found, or that the ultimate model of the atom has been discovered. The development of the mechanics both of planetary motion and of the atom has made our outlook sceptical. Even disregarding the statistical turn of quantum physics, we must therefore envisage all three possibilities of improving a given description.

When we always repeat the procedure of improving the description, going from $D^{(1)}$ to $D^{(2)}$ and then to more and more precise descriptions, we arrive at a sequence of convergent descriptions. Determinism can now be formulated as the hypothesis that in this sequence the probability of the prediction of the later state E increases toward the value 1, *and* that there exists an ultimate precise description D toward which the approximate descriptions converge. This hypothesis is expressed by the schema:

$$D^{(1)} \quad D^{(2)} \quad D^{(3)} \quad \cdots\cdots\cdots\cdots \to D$$
$$p^{(1)} \quad p^{(2)} \quad p^{(3)} \quad \cdots\cdots\cdots\cdots \to 1 \ . \tag{1}$$

In this schema, the description E of the volume v at the time t_2 remains unchanged. But determinism requires that, however precise we

---

[1] Ludwig Boltzmann, *Vorlesungen über Gastheorie*, Bd. II (Leipzig, 1898), p. 260. The following is an English translation of the passage: Since it is fashionable today to envisage a time when our philosophy of nature will have completely changed, I will mention the possibility that the fundamental equations of motion of the individual molecules may turn out to be mere approximations, referring to averages which, according to the calculus of probability, result from the combined action of many individuals that move independently and constitute the surrounding medium.

wish to make the description E, a convergent schema as formulated in (1) results.

Let us discuss this hypothesis in more detail. Even for classical physics, for which points 2 and 3 were regarded as settled, point 1 has always been a source of difficulties. If the universe is spatially infinite, the volume V would increase with every new description $D^{(i)}$, and the ultimate initial description D would have to express the state of the whole infinite space at a certain time t_1. We thus would need an infinite number of parameters to formulate D. (Incidentally, it would be irrelevant to what time point t this cross section through the universe is referred; "initial" means no more than "selected starting point of the causal chain".) But then it does not make much sense to speak of an ultimate description D, because an infinite number of independent parameters cannot be combined in a single state description.

Einstein's theory of relativity has modified this logical quandary in two ways. First, Einstein's cosmology has made it probable that the universe is spatially finite; this result would allow for the assertion of an ultimate description D as far as point 1 is concerned. But even if Einstein's supposition concerning a finite universe should turn out to be untenable, his theory supplies other reasons for regarding the relevant volume V as finite. It was pointed out with respect to figure 8, §5, that the light cone ending in the event A comprises all events that can ever causally contribute to A. Given a small volume v and a time t_2, we can therefore compute the largest volume V_1 in such a way that occurrences in V_1 at the time t_1 can causally contribute to the state of v at the time t_2. The occurrences outside the volume V_1 can be neglected, because it would take too long a time for causal effects spreading from them to reach v: such effects would arrive in v after the time t_2. Therefore it is sufficient if the ultimate description D refers to V_1; and we need not know all the parameters controlling an infinite spatial expansion.

This turn of the problem, however, also has its disadvantages. If the observer is situated in v, he cannot acquire before the time t_2 a total knowledge of the condition of V_1 existing at the time t_1. Any information traveling from the border of V_1 to the observer is restricted in its transmission to the speed of light. Thus the relativistic observer is essentially in the same predicament as the observer of classical physics: whereas the latter is unable to acquire at the time t_1 a knowledge of infinite space, the observer of the relativistic physics cannot acquire at the time t_1 the knowledge of that finite part of space which is required for predictions concerning an event at the time t_2.

The latter conclusion holds, at least, for an infinite relativistic universe. And we may add that for such a universe the size of the volume V_1 is dependent on the time difference $t_1 - t_2$; if this difference is increased, knowledge of a larger volume V_1 is required. In contrast, if Einstein's supposition concerning a spatially finite universe is correct, the total volume of space constitutes a maximum beyond which V_1 can never increase. It would then be physically possible to acquire at the time t_1 a knowledge of the state of the total universe at some earlier time t_0, even though this time t_0 might be billions of years ago. Once this early state is known, any future state could be predicted in a deterministic world. If the universe is spatially finite, therefore, point 1 would offer no logical difficulties for the existence of an ultimate description, either in the classical or in the relativistic world, inasmuch as this description would not require an infinite number of parameters.

There remains the question of the physical presuppositions which are included in the assumption that an ultimate description of the state of the universe is possible. If the observer is situated at some place x at the time t_1, we would have to assume that somehow all points of the universe emit, at the time t_0, signals arriving in x at the time t_1. The point event $\{x, t_1\}$ then would keep records, so to speak, of the total universe at the time t_0. Such a possibility appears highly improbable. But we need not go into an analysis of this question, because the spatial finitude of the universe is problematic. We shall rather attempt to formulate determinism in such a way that the formulation can be applied even to a spatially infinite world. The possibility of a finite space of enormous extension will then be included.

The following formulation, which has been discussed repeatedly, suggests itself. The ultimate description D of initial conditions refers only to a finite spatial volume v, but we require in addition that the boundaries of the volume v be known throughout the period from t_1 to t_2. This formulation may be illustrated by figure 16, in which the vertical lines express time, whereas the horizontal lines represent spatial volumes. The volume v extends, at the time t_1, from 1 to 2; at the time t_2, it is represented by the line from 3 to 4. The boundaries of v during the period from t_1 to t_2 are given by the lines 1-3 and 2-4. Since all influences from the spatial environment must pass through these boundaries, the infinite remainder of space is here replaced by a finite domain and thus is made describable in terms of a finite number of parameters, as far as the system considered is concerned. Likewise, the relativistic difficulty of acquiring information concerning the large volume V_1 is eliminated, because the volume v can be kept rather small. It is true that the description D, in the new interpretation, does not help us

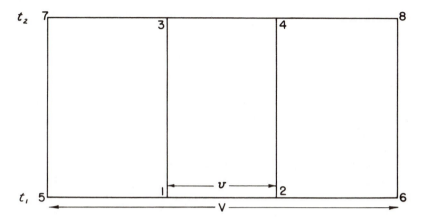

Fig. 16. Determination of the state of the spatial volume v at time t_2 in terms of the state of the spatial volume V at time t_1 and the occurrences at the boundary of V during the time from t_1 to t_2.

at the time t_1 to make predictions, because a knowledge of what happens later at the volume's boundaries can only be acquired at later times. But the new method allows us, at least, to formulate the hypothesis of determinism even for a spatially infinite universe, whether it be classical or relativistic.

Using this formulation, however, we encounter the following difficulty. When we refer to a rather small spatial volume v, it may be easy to predict its future states from simultaneous boundary conditions, but not from much earlier ones. Assume that we wish to predict the weather in a city. We select the area of the city as the basis of our initial volume v, and imagine a hemispherical surface in the sky fitting right on top of the city. This is the boundary of our system. To predict what the weather will be after ten years would be easy, if we knew the meteorological conditions on our boundary surface at the time of the forecast. In contrast, the initial conditions, concerning the weather in the city today, have become irrelevant. Our so-called weather prediction, however, would be not a long-range forecast, but merely an interpolation of meteorological data concerning the city between simultaneous data on the boundary surface. For us to make this interpolation a deterministic world would not be required.

To escape this consequence, we formulate the following additional requirement. Given a spatial volume v, we surround it by a larger volume V which we describe by $D^{(1)}$ at the time t_1 and which is indicated in figure 16 by the line 5–6. In order to predict the state E for the small volume v at a later time t_2, we include in $D^{(1)}$ the boundary conditions of the volume V during the time from t_1 to t_2. These boundaries are represented in figure 16 by the lines 5–7 and 6–8. When we say that the state E of v can be predicted with the probability $p^{(1)}$, we mean that such a prediction can be achieved however large the volume V that we select may be. For instance, when we say that the weather in the city (volume v) can be predicted for the next day with a probability of 80 per cent, we mean that this can be done if weather data concerning the whole continent and an imaginary hemispherical surface on top of it (boundary of volume V) are given today, and if the weather data at the boundary of this hemispherical shell up to tomorrow are known. The meteorologist, of course, would argue that this is not a weather forecast, because the additional data concerning the continent's border would not be available before tomorrow; and he would add that he could do very well without them, because storm fronts move slowly and the distant weather has no great influence on the weather of the city tomorrow. Furthermore, the imaginary hemisphere stretches into zones of the atmosphere which are irrelevant to the weather. All this is true; but the argument shows why we use this formulation. A prediction from a large volume V is practically equivalent to a prediction from an infinite volume, or to a prediction from the volume V_1 required for the relativistic universe, even if the volume V is much smaller than the volume V_1. The influence of the data at the boundary thus drops out, practically speaking. The new formulation has the advantage that it avoids a description of an infinite volume, or of the inaccessible volume V_1. It uses, instead, the description of a none-too-large volume V and of the conditions existing during a finite time at its boundary. For this reason, it allows us to formulate determinism.[2]

Applying these results to the schema (1), and including in the $D^{(i)}$ the boundary conditions at the volume V between t_1 and t_2, we define determinism as follows: However large we choose the volume V for a given volume v, if we keep V and v constant in the sequence of descriptions of schema (1), there exists an ultimate description D which predicts the state E of the volume v with the probability 1. In other

---

[2]The formulation of determinism by the use of a finite volume v and boundary conditions has been discussed repeatedly. It was given, for instance, by Moritz Schlick, in "Causality in Everyday Life and in Recent Science", *Univ. Calif. Publ. Philos.*, Vol. 15 (1932), pp. 99–126, reprinted in Herbert Feigl and Wilfrid Sellars (eds.), *Readings in Philosophical Analysis* (New York, Appleton-Century-Crofts, 1949), pp. 515–533. However, this presentation by Schlick does not mention the additional requirement concerning the larger volume V, without which the formulation is practically empty.

words, schema (1) stands for a class of schemata each of which results for a certain volume V.

Let us now formulate indeterminism, still disregarding quantum physics; we may speak here of *classical indeterminism*. In this conception, we still assume that the probability of prediction can be increased toward the value 1. But we deny the possibility of an ultimate description D; that is, we do not assume that the sequence of more and more precise descriptions converges toward a limit. This conception is expressed by the schema

$$D^{(1)} \quad D^{(2)} \quad D^{(3)} \quad \cdots \cdots \cdots \cdots \cdots$$
$$p^{(1)} \quad p^{(2)} \quad p^{(3)} \quad \cdots \cdots \cdots \cdots \cdots \to 1 \; . \tag{2}$$

If the descriptions $D^{(i)}$ are restricted to conditions at the time t_1, the infinity of space would be a sufficient reason for rejecting the existence of an ultimate description D, as explained. But schema (2) can also be applied to a spatially finite world, or to descriptions $D^{(i)}$ which include the boundary conditions of the volume V from t_1 to t_2. In the second instance, we maintain that schema (2) results, provided we choose the volume V sufficiently large in comparison to the volume v to which the prediction refers. This means that there is a volume V around v such that for all larger surrounding volumes, schema (2) results.

In discussing the form (2) it should be noted that this form is logically consistent. There is no way of inferring from the existence of a limit in the sequence of the $p^{(i)}$ that the sequence of the $D^{(i)}$ must also have a limit. To give an illustration, consider the two sequences

$$1, 2, 3, 4, \cdots \cdots \cdots$$
$$\frac{1}{1}, \frac{1}{2}, \frac{1}{3}, \frac{1}{4}, \cdots \cdots \cdots \tag{3}$$

Although these two sequences are in one-one correspondence, only the second has a limit, the value 0, whereas the first has no finite limit.

This does not prove, of course, that the sequence of $D^{(i)}$ has no limit. Such a proof cannot be given on logical grounds; we have to refer to arguments from physics. However, in denying the existence of an ultimate description D, we may refer to point 2 and question the assumption that the number of inner parameters of the system is finite. And we may argue with reference to point 3 that the simple mathematical relationships known under the name of equations of motion are only approximately valid, and that, further research will replace them by much more complicated relationships, a process of improvement which need not converge to an ultimate formula. We have thus formulated a conception denying determinism; and in the formulation referring to boundary conditions, the rejection of an ultimate description D is not derived from the infinity of space, but is based on other grounds.

Both determinism and indeterminism, obviously, are hypotheses, are extrapolations going beyond the observables. What evidence can be adduced in favor of one or the other?

Since we are concerned here with hypotheses concerning the limit of a sequence, and the limit itself can never be observed, we must try to construct evidence from convergence properties of the sequence. From this viewpoint, the following consideration suggests itself. The convergence of probabilities of prediction may be regarded as well supported by evidence and inductive inference. But it is usually forgotten that the same kind of evidence reveals a divergence of probabilities as soon as longer time intervals are considered. Let us first study this divergence by the use of the formulation that does not refer to boundary conditions. Given a description $D_1^{(1)}$ of initial conditions in v at the time t_1, we may be able to predict the future state E_2 of v at the time t_2 with a high probability $p_2^{(1)}$; but if we wish to know the state E_3 of v at a later time t_3, the same description will attain this aim only with a lower probability $p_3^{(1)}$. It is true that $p_3^{(1)}$ can be improved by the use of a more precise description $D_1^{(2)}$ at the time t_1. But for $D^{(2)}$, again, the predictive probability will decrease when we wish to make predictions for a time t_4 which is later than t_3. We thus have a double process of convergence and divergence, which may be expressed in the following schema, in which the subscripts indicate the time referred to:

$$
\begin{array}{ccccc}
D_1^{(1)} & D_1^{(2)} & D_1^{(3)} & \ldots & D_1^{(i)} \ldots \\
p_2^{(1)} & p_2^{(2)} & p_2^{(3)} & \ldots & p_2^{(i)} \ldots \to 1 \\
p_3^{(1)} & p_3^{(2)} & p_3^{(3)} & \ldots & p_3^{(i)} \ldots \to 1 \\
\cdots & \cdots & \cdots & \cdots & \cdots \\
p_k^{(1)} & p_k^{(2)} & p_k^{(3)} & \ldots & p_k^{(i)} \ldots \to 1 \\
\cdots & \cdots & \cdots & \cdots & \cdots \\
\downarrow & \downarrow & \downarrow & & \downarrow \\
p & p & p & \ldots & p \ldots
\end{array}
\qquad (4)
$$

In this lattice arrangement, the probabilities of all horizontal rows converge to 1, but those of vertical columns converge to the over-all probability (or antecedent probability) p of the state E. This means that, if we can make any probability statement at all about the occurrence of E after a long time, we must use our knowledge of the probability of E in general, and cannot base our predictions on the initial state $D_1^{(i)}$; therefore this probability p is the same for all $D_1^{(i)}$, and thus for all columns. For instance, when we are asked to predict the maximum temperature for a certain day in July in a certain city after one hundred years, we shall use the average probability resulting from general statistics for July taken for this city; our knowledge of weather conditions today is then quite irrelevant. Since for columns farther to the right the first probabilities $p_2^{(i)}$, $p_3^{(i)}$, and so on, are rather high, the probability values decrease in these columns toward the value p. This is what we mean by the divergence of these probabilities.

We must now extend the divergence to the formulation referring to boundary conditions. We recall that here the schema (4) stands for a class of such schemata, each resulting for a certain volume V, whereas the volume v to which the prediction refers remains constant. Each $D_1^{(i)}$ includes a description of the conditions at the boundaries of V during the period from t_1 to t_k. When we extend the period of the forecast from the time t_k to a later time t_m, the description $D_1^{(i)}$ is to be supplemented by statements about the conditions at the boundary of V during the period from t_k to t_m; we will assume that these additional data are stated by the use of the same kind of parameters and to the same degree of exactness as were employed for the description of such data up to the time t_k. When we proceed along a column, therefore, the term "$D_1^{(i)}$" changes its meaning continually, but in a well-defined way; in fact, it stands for a set $\sigma^{(i)}$ of individual descriptions each of which belongs to a certain t_k, or t_m, according to the rules given. We can thus speak meaningfully of the probabilities $p_k^{(i)}$ and $p_m^{(i)}$.

Now if V is very small, or even identical with v, the vertical columns need not lead to a divergence of probabilities, that is, $p_m^{(i)}$ need not be smaller than $p_k^{(i)}$. We may refer to the illustration in which the weather in a city was predicted and the boundary surface was defined by a hemispherical shell on top of the city. However, if V is sufficiently large, we shall find the divergence expressed in schema (4), because the conditions at the boundary of V are practically irrelevant to the occurrences within v. This means: there is a volume V such that for it and every larger volume, schema (4) results. Using again the example of the attempt to predict what the maximum temperature on a July day in a certain city one hundred years hence will be, we may choose as boundary conditions, as above, meteorological data on a hemispherical surface of a radius of five thousand miles around the city. These data

will not help us very much; and all we can do is predict the temperature with the over-all probability p. Thus the convergence to p may
again be asserted for the columns. The reasons for the loss of predictive power are given in points 2 and 3 above.

Schema (4) represents a lattice of *nonuniform convergence*. Given an
exactness ϵ and a time t_k, we can find a description $D_1^{(i)}$ at time t_1 such
that it predicts the state of v at time t_k with a probability $p_k^{(i)} \geq 1 - \epsilon$.
But we then can find a time $t_m > t_k$ such that the same description $D_1^{(i)}$
predicts the state of v at t_m only with a lower probability $p_m^{(i)}$. For
small values ϵ, there is no description $D_1^{(i)}$ which predicts the state of
v for all times with a probability equal to or greater than $1 - \epsilon$.

Let us use these results for the discussion of the existence of an
ultimate description. We face here a peculiar difficulty. If a sequence
of descriptions is given by some rule, we may be able to find out
whether it possesses a limit, that is, whether it converges to an ultimate description. For instance, an infinite sequence of little circles
contracting toward a point in a plane may be regarded as a sequence of
approximate descriptions of the point; this sequence has a limit in an
ultimate description, namely, the coördinates of the point. However,
the sequence of descriptions of physical states is not given by some
defining rule, and we are confronted by the problem of judging by means
of inductive inferences whether the sequence has a limit. We must
therefore ask what kind of inductive evidence can be regarded as
supporting the hypothesis that the sequence has a limit.

Now it appears plausible to contend that, if schema (4) showed a
uniform convergence, this fact could be regarded as supporting evidence for the existence of a limit. Every column of (4) would then
converge to some value $p^{(i)}$ such that the last row, which contains all
the $p^{(i)}$, would converge toward 1. We then might conclude that there
is an ultimate description D for which the value of the predictive probability is equal to 1 throughout. Such a last column, in fact, would be
uniquely determined by all the other columns as their limiting case.
For a spatially infinite universe, for which we have to refer to boundary conditions, the definition of D is a little complicated. Here D
represents an infinite set σ of individual descriptions, each corresponding to a specific time t_k and supplying the value $p_k = 1$; since
the set σ of descriptions is well defined, as explained above, the
existence of this set represents a meaningful hypothesis.

In contrast, if schema (4) shows a nonuniform convergence, an
inductive inference concerning a last column cannot be made in a
unique way. Proceeding by inferences along each row, we would put a
probability of the value 1 into every place of such an ultimate column,

as indicated in schema (4). But proceeding by inferences referring to columns as wholes, we would conclude that the probabilities in the ultimate column should converge toward p, and thus cannot have the value 1 throughout. In other words, the right lower corner of the schema cannot be filled out in a consistent way: if we come from the top, going along the last column, we would put the value 1 into this field; if we come from the left, going along the last row, we would put the value p into it. This kind of ambiguity is familiar from the theory of limits; a double transition to a limit often leads to different results, since the result depends on the order in which the two transitions are performed. This means that for nonuniform convergence the schema does not define an ultimate column in a consistent way; and therefore the schema cannot be regarded as inductive evidence for the existence of such a column.

We conclude that a nonuniform convergence of schema (4) must be regarded as absence of supporting evidence for the existence of an ultimate description. This does not mean that the schema represents evidence against the assumption of a limit; it merely does not support this assumption. It makes the assumption of a limiting description appear as an empty addition, which does not manifest itself in observable relationships.

Let us investigate this problem by studying an example in which a nonuniform convergence is combined with the existence of an ultimate description—an example, therefore, which shows that this combination is not logically contradictory. Such an example is presented by the ergodic hypothesis of the classical kinetic theory. At first glance one might conclude that Liouville's theorem, according to which any domain in the phase space is mapped by the ergodic path on a domain of equal size (see fig. 13, in §10), leads to the result that the inexactness of a description must forever remain within a small interval ϵ. But this conclusion would be incorrect. Liouville's theorem does not require that the domains keep their shapes. If A is a spherical domain (or, more precisely, a hyperspherical domain), then the domain A' onto which it is mapped by the path may have a starfish-like shape (see fig. 17), though it covers an area of the same size as A.[3] The thin branches of this extremely ramified domain A' will eventually extend over the whole space. Therefore, if the initial position of the phase point is known within a small interval ϵ, we cannot conclude that this knowledge determines the position of the point for all times within a small interval δ of exactness which is a function of ϵ. In fact, with

[3]In the ergodic measure, if the energy surface is considered; see the explanation in §10, n. 8, of the present study.

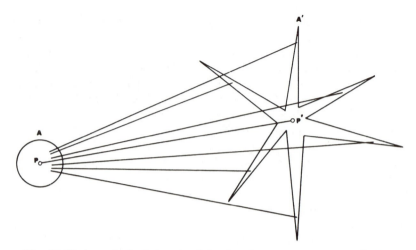

Fig. 17. If the spherical domain A in the phase space is mapped, by the paths of the phase points, onto the domain A', the latter domain may have a starfish-like shape the arms of which extend over a wide area.

growing time we may expect to find the point at any distance larger than a given δ from the predicted place P', if we define P' as the point on which the point P in the center of A is mapped. The ergodic determinism therefore admits of a nonuniform convergence: although there exists an ultimate description that makes strict predictions possible, every approximation to it loses in predictive power with growing time, and the resulting schema displays a nonuniform convergence.

In this analysis of the ergodic hypothesis we have first assumed determinism and then studied convergence properties. When we proceed inversely, using the same example, and study convergence properties first, we would find no inductive evidence for determinism. If the domain A' extends in starfish shape over a large part of the phase space, we have no way of distinguishing by means of measurements between points inside one of the thin branches of A' and points outside it, because the exactness of observations is limited. If only the time distance t between A and A' is large enough, the starfish domain A' has all the properties of a well-rounded domain encircling A', at least as far as empirical observations are concerned. The ergodic determinism appears as a perfect imitation of indeterminism, so perfect that the imitation cannot be distinguished from the original.

This is what was meant above when we said that a schema of non-uniform convergence must be regarded as absence of inductive evidence for the existence of an ultimate description. And the dual convergence of schema (4) represents the outcome of all observational evidence in classical physics. Judged in terms of observable convergence properties, the hypothesis of an ultimate description appears, therefore, as an empty addition, which can very well be dispensed with. And it is hard to see what kind of empirical evidence could be adduced for determinism other than evidence derived from convergence properties of approximations. If it is claimed that classical physics proves determinism, the claim has sprung from admiration of the success of mathematical methods in physics rather than from a sober examination of observational evidence. And if it is argued that classical physics presupposes determinism, we must answer that there is no chapter of physics in which determinism is used as an indispensable presupposition. Only convergence properties are presupposed; but they can be satisfied just as well if there exists no ultimate description. Determinism is logically compatible with classical physics; but determinism is by no means proved by it, nor made probable by inductive evidence. Even for classical physics, determinism is a redundant addition to the system of hypotheses which formulate the basic laws of the physical world.[4]

This is the answer which an analysis of meanings gives to the question of determinism in kinetic theory. The ergodic path can be regarded as a good approximation to observable relationships, but it need not be interpreted as being more than that. The fact that it is possible to derive a probability metric for the evolution of closed systems from the causal laws of mechanics is a result of great importance, which can be taken over into an indeterministic conception because here the results of the ergodic hypothesis remain approximately valid. Even if physical laws are regarded as not absolutely strict, a probability metric can be derived from them; that is, laws of lower probability can be derived from other laws holding with a very high probability. But the conclusion that the evolution of closed systems is governed by strict rules is rejected. There is no reason for accepting this deterministic interpretation of thermodynamical statistics. We are allowed to regard even the evolution of the universe as a genuine probability

---

[4]The criticism of determinism within the frame of classical physics, presented in this section, was developed in some of my earlier papers, published in part before the discovery of quantum-mechanical indeterminacy. See Hans Reichenbach, "Die Kausalstruktur der Welt und der Unterschied von Vergangenheit und Zukunft", *Sitzungsber, d. Bayer. Akad. d. Wiss.*, math.-naturwiss. Abt., 1925, pp. 135-138; "The Principle of Causality and the Possibility of Its Empirical Confirmation" in *Modern Philosophy of Science* (London, Routledge & Kegan Paul and New York, Humanities Press, 1959), written in 1923 and published in *Erkenntnis*, Bd. 3 (1932), pp. 32-64; and "Stetige Wahrscheinlichkeitsfolgen", *Zeitschr. f. Physik*, Bd. 53 (1929), pp. 274-307.

chain, in which each successive step is a matter of chance, whereas the sequence as a whole is governed by statistical laws as dependable as the so-called strict causal laws of physics.

● *12. The Probability Lattice*

The mathematical and logical methods for the analysis of probability relations have been developed in the theory of probability. A short account of these methods and the notation to be used may be inserted before we turn to applying these methods to the study of mixing processes.[1]

A probability is a relation between two classes, considered with respect to a sequence. For instance, the probability that a man of twenty-one years will die within a year from tuberculosis, relates two classes: the *reference class A*, that is, the class of men of twenty-one years; and the *attribute class B*, the class of persons dying of tuberculosis. Such a probability is written in the form

$$P(A, B) = p \; .\tag{1}$$

The value of this probability is found by counting the relative frequency of B within A; this is done with respect to a certain sequence of deaths in a certain country occurring within a certain period.

However, it is not directly the relative frequency of observed events which we regard as the interpretation of the probability p. We assume that this relative frequency, for future observations, will converge toward a limit; and p is measured by this limit of the frequency. Let $N^n(A)$ be the number of elements belonging to A, and $N^n(A.B)$, the number of elements belonging both to A and B, counted up to the n-th element of the sequence. Then the limit of the relative frequency is given by

$$\lim_{n \to \infty} \frac{N^n(A.B)}{N^n(A)} \; .\tag{2}$$

This expression formulates the frequency interpretation of (1).

That such a limit exists and will be observed in future continuations of such observations is an assumption which we make when we say that the probability (1) exists. It is, of course, sufficient that the sequence has a *practical limit*, since all practical probability sequences are finite and thus cannot supply a convergence in the mathematical

[1] I use the symbolic notation developed in *ThP*. For the meaning of (1) see *ThP*, pp. 47 and 50.

sense of the term "limit". By a practical limit we understand a convergence of a relative frequency which rather early reaches its final value within a small interval of exactness and stays within this interval for the rest of the time. These brief remarks may suffice here; for a detailed discussion of the limit assumption, I refer the reader to another publication.[2]

We are often concerned with probabilities in which the general reference class A is identical with the class of events of the sequence; then it is convenient to omit "A" in the probability expression, and instead use an absolute notation.[3] For instance, the probability of good-weather days is usually counted in the sequence of days; so we can omit the symbol of the reference class and write this probability in the form

$$P(B) \ , \tag{3}$$

where "B" means "good-weather day". Only when special reference classes are used will this be expressed in the symbolization. For instance, the probability that a cold day is a good-weather day would then be written in the form (1), and A would be the class of cold days.

If we wish to characterize the inner structure of a probability sequence, we use, among other devices, the probability influence of predecessors in the sequence, that is, we investigate the existence of aftereffect. A random sequence has no aftereffect; for instance, the probability of throwing a 6 with a die is the same whether or not the preceding throw resulted in a 6. There are further properties which a sequence must satisfy in order to be called a random sequence. A general theory of probability, however, is not restricted to random sequences. It includes a chapter on a *theory of order*, which treats sequences of various types of order and defines special types. The random sequences constitute merely one of these types, whereas other types of sequences can be equally important for physical applications. It was pointed out above that the ergodic sequence has an aftereffect. Another illustration is supplied by weather observations, as mentioned above. This sequence has an aftereffect, because a day of good weather is usually followed by a similar day, and a day of bad weather by a bad-weather day. But the aftereffect subsides: For the second day after a day of good weather, the probability of good weather has become smaller; and when we wish to compute the probability of good weather twenty days after a good-weather day, we can forget about the aftereffect and use the over-all probability of good weather for the computation.

---

[2]*Ibid.*, §§16, 64–66, and 91.

[3]*Ibid.*, p. 106.

An aftereffect probability is written by means of *phase superscripts.* For instance, if again B is the class of good-weather days, the expression

$$P(B, B^1) \tag{4}$$

denotes the probability that, in the sequence of days, a day of good weather is followed by a day of good weather. The general reference class of days is here omitted as understood. This probability is higher than (3), that is, higher than the over-all probability of good weather, because of the probability aftereffect. The frequency interpretation of (4) is given by the expression

$$\lim_{n \to \infty} \frac{N^n(B.B^1)}{N^n(B)} . \tag{5}$$

This means that we count the number of pairs of good-weather days among the number of good-weather days.

Aftereffect plays an important part in the diffusion of gases. Suppose that we have a container divided into two parts by a partition (fig. 18); into the left part, to be called "B", we put nitrogen, and into the right one, called "$\bar{B}$", oxygen. When we now take out the dividing wall, a mixing process begins. Every molecule travels in a zigzag path through the container; and after some time, molecules of nitrogen and molecules

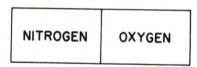

Fig. 18. Two gases separated by a partition. When the partition is taken out, a mixing process begins.

of oxygen will be in equal numbers in B and in $\bar{B}$. But the path of a molecule shows aftereffect. Once it is in B, it will probably still be in B a short time later; and the same applies to its stay in $\bar{B}$. When we write down these sequences for all the molecules participating in the diffusion process, we obtain a probability lattice:

$$
\begin{array}{l}
x_{11} \; x_{12} \; \bullet \bullet \bullet \; x_{1i} \; \bullet \bullet \bullet \to p \\
x_{21} \; x_{22} \; \bullet \bullet \bullet \; x_{2i} \; \bullet \bullet \bullet \to p \\
\bullet \bullet \bullet \bullet \bullet \bullet \bullet \bullet \bullet \bullet \bullet \\
x_{k1} \; x_{k2} \; \bullet \bullet \bullet \; x_{ki} \; \bullet \bullet \bullet \to p \\
\bullet \bullet \bullet \bullet \bullet \bullet \bullet \bullet \bullet \bullet \bullet \\
\downarrow \quad \downarrow \quad \quad \downarrow \\
p_1 \; p_2 \; \bullet \bullet \bullet \; p_i \; \bullet \bullet \bullet \to p
\end{array} \tag{6}
$$

Each horizontal row represents the history of a molecule, cut into events by short time intervals Δt. We thus avoid dealing with continuous probability sequences.[4]

To simplify the presentation, we will first treat the lattice with reference to the nitrogen molecules only, omitting the oxygen molecules from the lattice. Since in the beginning of the mixing process all nitrogen molecules are in B, the first vertical column has a "B" in every place.[5] The over-all probability, or absolute probability, of B in a row[6] is $p = 1/2$; it is written in the form

$$P\left(B^{ki}\right)^i = p \ . \tag{7}$$

This value holds for any row k; we indicate this fact by using "k" as a free variable. The superscripts indicate the position of the event in the lattice. Repetition of a superscript outside the parentheses indicates the running superscript, or bound variable; in other words, this repetition indicates the index of summation in the frequency interpretation and thus the direction of counting. The expression (7) is therefore a *horizontal probability.*

We can also construct a *vertical probability* and write for the absolute probability of B in a column i the value

$$P\left(B^{ki}\right)^k = p_i \ . \tag{8}$$

For the first column we have $p_1 = 1$, because it contains only B-terms. The second column will contain a few $\bar{B}$-terms; and gradually the p_i will thus become smaller and converge toward p. That is, we have the relations

$$P\left(B^{ki}\right)^k > P\left(B^{ki}\right)^i \ , \tag{9}$$

$$\lim_{i \to \infty} P\left(B^{ki}\right)^k = P\left(B^{ki}\right)^i \ . \tag{10}$$

The convergence relation (10) between vertical and horizontal probabilities constitutes the mathematical expression of the mixing process.

The aftereffect existing for the rows can be written in the following form, when we indicate the phase difference by putting "-1" into the superscript of the reference term:

---

[4]But a lattice presentation in terms of such sequences could also be given; see *ibid.*, p. 247.

[5]["B" and "$\bar{B}$" are now used to denote the event that at a given time a given molecule is in compartment B or compartment $\bar{B}$.—M. R.]

[6][The phrase "the probability of B (or $\bar{B}$) in a row (or column)" is a harmless ellipsis. It may be understood either to mean "the probability of 'B' (or '$\bar{B}$') in a row (or column)" or to mean "the probability of the event B (or $\bar{B}$) in the sequence of events represented by a row (or column)". The same results follow, whichever interpretation is selected.—M. R.]

$$P\left(B^{k,i-1}, B^{ki}\right)^i > P\left(B^{ki}\right)^i ,$$

$$P\left(\overline{B}^{k,i-1}, B^{ki}\right)^i < P\left(B^{ki}\right)^i . \tag{11}$$

These relations express a positive aftereffect, that is, a tendency to stay. A negative aftereffect, that is, a tendency to change, exists in other types of sequences, but does not occur in the rows of a lattice of mixture. There are simplifying rules for the use of superscripts: a constant can be added or subtracted in the superscript, provided this is done in all terms alike.[7] For instance, the left-hand side of the first relation (11) can be written in the form $P(B^{ki}, B^{k,i+1})^i$. The aftereffect expressed in (11) becomes smaller with growing distance; we have for any integer $r > 0$:

$$P\left(B^{k,i-r-1}, B^{ki}\right)^i < P\left(B^{k,i-r}, B^{ki}\right)^i . \tag{12}$$

When we go to infinite distances, the aftereffect subsides completely, and the phase probability converges to the absolute probability:

$$\lim_{r \to \infty} P\left(B^{k,i-r}, B^{ki}\right)^i = P\left(B^{ki}\right)^i = p ,$$

$$\lim_{r \to \infty} P\left(\overline{B}^{k,i-r}, B^{ki}\right)^i = P\left(B^{ki}\right)^i = p . \tag{13}$$

A practical equality will already be reached for large r.

In contrast, the columns show no aftereffect, because the rows are independent of one another; that is, we have

$$P\left(B^{k-1,i}, B^{ki}\right)^i = P\left(B^{ki}\right)^k . \tag{14}$$

Relation (14) refers to vertical probabilities; relation (10) connects vertical and horizontal probabilities. We may ask whether these relations are derivable from relations (11)--(13), which refer to horizontal probabilities alone. Such a derivation cannot be given wholly in terms of the general laws of probability, because horizontal and vertical probabilities are, in general, mutually independent mathematical quantities; but it can be given if the lattice satisfies two restrictive conditions, which we must now explain.

The first restrictive condition expresses the *independence* of the rows and refers to horizontal probabilities. It states that the probability of B in a row, and any phase probability also, remains unchanged in a subsequence selected by reference to the occurrence of B in some other row. For instance, we assert that

$$P\left(B^{k,i-1} . B^{k-1,i}, B^{ki}\right)^i = P\left(B^{k,i-1}, B^{ki}\right)^i . \tag{15}$$

---

[7] See *ThP*, p. 137.

This means that the aftereffect probability for B, referred to the immediate predecessor (right-hand side), is not influenced by the occurrence of a term "B" in the preceding row at the same place (term "$B^{k-1,i}$" on the left-hand side). When we wish to formulate this condition for any group of predecessors and any terms occurring in other rows, it is convenient to use a disjunction "$B_1 \vee \ldots \vee B_m$"; we then need not write separate conditions for B and $\bar{B}$. We thus arrive at the following formulation for the condition of independence:

$$P\left(B_{m_r}^{k,i-a_r} \ldots B_{m_1}^{k,i-a_1} . B_{n_1}^{k+b_1,i+c_1} \ldots B_{n_s}^{k+b_s,i+c_s}, B_{m_0}^{ki}\right)^i$$

$$= P\left(B_{m_r}^{k,i-a_r} \ldots B_{m_1}^{k,i-a_1}, B_{m_0}^{ki}\right)^i , \qquad (16)$$

$$a_1 \ldots a_r > 0 , \qquad b_1 \ldots b_s \neq 0 .$$

The use of subsubscripts enables us to indicate the correspondence between subscripts and superscripts. A subscript having a subscript is a variable like any simple subscript. For instance, the subscript "m_r," may be equal to 1; then the term "B_{m_r}" assumes the value B_1. The condition $b_1 \ldots b_s \neq 0$ serves to put the corresponding B-terms into rows different from the k-th row to which the probability on the right is referred. The $c_1 \ldots c_s$ can be equal to 0. The condition that $a_1 \ldots a_r > 0$ restricts the terms in the reference class to predecessors of the term in the attribute class. If $r = 0$, all predecessors drop out, and condition (16) then refers to absolute probabilities. Likewise, if $s = 0$, all the terms referring to different rows on the left drop out, and the condition is then a trivial identity.

The second restrictive condition expresses a property called *lattice invariance*.[8] It connects horizontal and vertical probabilities. For instance, we have here

$$P\left(B^{k,i-1}, B^{ki}\right)^i = P\left(B^{k,i-1}, B^{ki}\right)^k . \qquad (17)$$

This means that the horizontal probability of finding a "B" after a "B" in the k-th row (left-hand side) reappears as a vertical probability referred to two consecutive columns: it measures the number of B-terms selected from the i-th column by the condition that the preceding column has a "B" at the same place (right-hand side). This condition is to be extended to any phase probability; however, it does not apply to absolute probabilities, since the probabilities for "B" in rows and columns are different. Furthermore, the condition is to include reference to terms in other rows, as in (16). The general formulation of

[8] *Ibid.*, p. 172.

lattice invariance then assumes the following form, in which the left-hand side is identical with that of (16), and the right-hand side results merely by changing the repeated superscript outside the parentheses:

$$P\left(B_{m_r}^{k,\,i-a_r} \ldots B_{m_1}^{k,\,i-a_1} \cdot B_{n_1}^{k+b_1,\,i+c_1} \ldots B_{n_s}^{k+b_s,\,i+c_s}, B_{m_0}^{ki}\right)^i$$

$$= P\left(B_{m_r}^{k,\,i-a_r} \ldots B_{m_1}^{k,\,i-a_1} \cdot B_{n_1}^{k+b_1,\,i+c_1} \ldots B_{n_s}^{k+b_s,\,i+c_s}, B_{m_0}^{ki}\right)^k , \qquad (18)$$

$$a_1 \ldots a_r > 0 , \quad r > 0 , \quad b_1 \ldots b_s \neq 0 .$$

To illustrate the meaning of this expression, it may be noted that the special form (17) results from (18) for $r = 1$, $a_1 = 1$, $s = 0$, $m_1 = 1$, $m_0 = 1$, if we put "B_1" for "B". The condition $r > 0$ excludes absolute probabilities from (18). The relation of lattice invariance, formulated in (18), can therefore be paraphrased as follows: Every phase probability counted horizontally is equal to the corresponding kind of vertical phase probability; and this applies likewise if reference to other rows is included.

Relation (14) can now be derived as follows. In (18) we put $s = 1$, $b_1 = -1$, $c_1 = 0$, $r = 1$, $a_1 = j-1$. On the right-hand side, we put $i = j$. Relation (18) then assumes the form

$$P\left(B_{m_1}^{k,\,i-j+1} \cdot B_{n_1}^{k-1,\,i}, B_{m_0}^{ki}\right)^i = P\left(B_{m_1}^{kl} \cdot B_{n_1}^{k-1,\,j}, B_{m_0}^{k,j}\right)^k . \qquad (19)$$

Because of the independence condition (16), the term "$B_{n_1}^{k-1,\,i}$" on the left can be dropped. The term on the left in (19) can be regarded as resulting from the left side of (18) when we put $s = 0$, $r = 1$; and, according to (18) this expression equals the expression resulting from the right-hand side of (19) when the term "$B_{n_1}^{k-1,\,j}$" is canceled. We thus find that canceling the latter term leaves the right-hand side unchanged, and we have:

$$P\left(B_{m_1}^{kl} \cdot B_{n_1}^{k-1,\,j}, B_{m_0}^{kj}\right)^k = P\left(B_{m_1}^{kl}, B_{m_0}^{kj}\right)^k . \qquad (20)$$

Since we have in the first column only B-terms, the first term in the expressions on either side of (20) assumes the form "B^{kl}" for every row, and thus can be dropped. Relation (20) is then identical with (14), if we put $j = i$.

Furthermore, relation (10) is derivable as follows. In (18) we put $s = 0$, $r = 1$, $a_1 = j-1$; and, as in (19), we put $i = j$ on the right side. Omitting the subscripts, that is, writing the result only for B-terms, we thus find

$$P\left(B^{k,\,i-j+1}, B^{ki}\right)^i = P\left(B^{kl}, B^{k,j}\right)^k = P_j . \qquad (21)$$

Since every row begins with a B-term, the last expression furnishes the absolute probability of B in the column j. That is, the value p_j of (21) corresponds to the definition (8). Relation (21) thus says that the absolute probability of B in the column j is equal to the horizontal aftereffect probability that a B is followed by a B at a distance of $j-1$ elements. According to (13), the first expression of (21) converges to the value p with growing j. The same must hold for the last expression; and thus (10) is proved.

This consideration shows: The mixing process is derivable for a lattice in which the rows possess an aftereffect and the first column is ordered—that is, not yet mixed—if the lattice satisfies the conditions of independence and lattice invariance. A lattice of this kind, which is characterized by relations (11)–(13), (16), and (18), will be called a *lattice of mixture*.

The given derivations show, in particular, the significance of the condition of lattice invariance. This condition allows us to transform horizontal probabilities into vertical probabilities. Since a row refers to the temporal succession of states of one molecule, it is often called a *time ensemble;* in contrast, a column is called a *space ensemble*, because it refers to the simultaneous states of many molecules. Relation (18) therefore formulates the condition which allows us to make an *inference from the time ensemble to the space ensemble*. This inference says that a cross section through the world lines of many individuals reflects the distribution within the world line of one individual. In the lattice of mixture, this inference applies directly to phase probabilities only; with respect to absolute probabilities, it holds only for later columns, after the mixing process has gone on for some time. In these columns, the frequency of B is practically the same as in the rows. However, the aftereffect structure of the rows is not transferred to columns, since the columns show no aftereffect, as has been stated earlier (14).

The lattice of mixture can be given other forms by selecting different distributions for the first column, whereas the general conditions (11)–(13), (16), and (18) remain satisfied. For instance, we can include both the nitrogen and the oxygen molecules in the lattice presentation by regarding alternate rows as representing the world lines of nitrogen and of oxygen molecules. We then have equal numbers of elements B and $\bar{B}$ in every column and thus find, throughout,

$$P\left(B^{ki}\right)^i = P\left(B^{ki}\right)^k \quad . \tag{22}$$

This means that the inference from time ensemble to space ensemble, applied to absolute probabilities, holds for every column, including the initial ones.

In contrast, relation (14) is not valid for this lattice, and we observe a negative aftereffect in columns. The first column consists of terms "B" and "$\bar{B}$" in strict alternation. As a consequence, the following columns still show a tendency to alternate and thus possess a larger number of pairs "$B.\bar{B}$" and a smaller number of pairs "$B.B$". This fact is formulated by the relations:

$$P\left(\bar{B}^{k-1,i}, B^{ki}\right)^k > P\left(B^{ki}\right)^k \quad , \qquad P\left(B^{k-1,i}, \bar{B}^{ki}\right)^k > P\left(\bar{B}^{ki}\right)^k \quad , \qquad (23)$$

$$\lim_{i \to \infty} P\left(\bar{B}^{k-1,i}, B^{ki}\right)^k = P\left(B^{ki}\right)^k \quad , \qquad \lim_{i \to \infty} P\left(B^{k-1,i}, \bar{B}^{ki}\right)^k = P\left(\bar{B}^{ki}\right)^k \quad . \quad (24)$$

Note that these relations do not contradict. (20); the latter relation expresses the independence of the rows for vertical counting and applies to the present lattice, too, since it is derived from (16) and (18). The equality sign in (20) is due to the term "$B_{m_1}^{k1}$" in the reference class; with respect to the first element "B", or "$\bar{B}$", of a row, an element "B" is independent of its predecessor in the column. Only when the first element of a row is not included in the reference term does aftereffect occur. This fact shows that the aftereffect in columns originates merely from the strictly alternating distribution in the first column, but not from a physical dependence between the elements.

This lattice can be regarded as a superposition of two lattices of the preceding type; or rather, as an interposition of one such lattice between the rows of another one. We thus derive, corresponding to (21):

$$P\left(\bar{B}^{k,i-j+1}, B^{ki}\right)^i = P\left(\bar{B}^{k1}, B^{kj}\right)^k = q_j \quad . \quad (25)$$

Using lattice invariance and the convergence relations (13) for horizontal probabilities, we derive

$$\lim_{j \to \infty} p_j = p = \lim_{j \to \infty} q_j \quad . \quad (26)$$

Every column j of the total lattice represents a Poisson sequence[9] controlled by the alternating probabilities p_j and q_j. The existence of aftereffect as formulated in (23)–(24) is shown in the theory of such sequences. With growing j—for later columns—the probabilities p_j and q_j become approximately equal, because of the convergence expressed in (26), and the column resembles more and more a sequence controlled by the one probability p. However, the convergence relation (10) is here replaced by the stronger relation (22), because we assumed that $p = 1/2$; this is shown in the theory of Poisson sequences, but is also intuitively clear.

It is an interesting fact that in this lattice the mixing process manifests itself not in a change of the probabilities $P(B^{ki})^k$ from column to column, but in the existence of aftereffect within each of the initial

[9] ThP, p. 296.

columns. Only for later columns does this aftereffect disappear. This is expressed in relations (23)–(24). We thus have here a space ensemble which shows, through internal dependence relations, that it is a passing stage in a mixing process.

Considering other forms of distribution for B in the first column of the lattice, we may ask what kind of relations are derivable for the general case. Relations (21) and (25) hold always; and since we can always regard the lattice as a superposition of two lattices, given by the first elements "B", or "$\bar{B}$", respectively, relation (26) leads again to the validity of relation (10). We thus find that every lattice of mixture satisfies relation (10). The probabilities for B in the columns converge always to the over-all probability of B in the rows. The stronger condition (22), according to which this over-all probability directly controls every column, holds only for the special case that the distribution of B in the first column equals the distribution of B in the rows. The proof of (22) for this case is easily given and need not be presented here.

The convergence of the column probabilities, expressed in relation (10), is therefore a general characteristic of all mixing processes, if we define a mixing process by relations (11)–(13), (16), and (18). For later columns, the inference from the time ensemble to the space ensemble is thus valid for every lattice of mixture. The only difference between rows and later columns is that the rows possess aftereffect, whereas the later columns do not. These results, which have here been derived for a disjunction of two terms "B" and "$\bar{B}$", are easily extended to a disjunction of m terms "B_1" ... "B_m".

In the history of the kinetic theory—in particular, in the work of Gibbs—the inference from time ensemble to space ensemble plays an important part, and some remarks about its logical nature may therefore be added. One often reads in the presentations of statistical mechanics that the inference is derivable within the calculus of probability. Such a statement is incorrect. The condition of lattice invariance is not derivable from the axioms of probability, but represents a specific condition which holds only for certain types of lattices. Whether this condition is satisfied for the physical processes considered is a matter of empirical knowledge.[10]

To study these relationships, let us investigate a simplified form of the inference, resulting for a lattice in which the rows have no aftereffect. It is easily seen that here the condition of lattice invariance, relation (18), leads to the conclusion that all columns, except perhaps the first, have a probability q for B which equals the probability of B

---

[10] See also the discussion of this problem in *ibid.*, pp. 174 and 280.

in the rows. In other words, relation (22) is here valid, and we have $q = p$.

Let us see whether this result is derivable from the independence condition. We consider only the n first rows. If the rows are independent, the possible arrangements of B and $\bar{B}$ in a column of length n can be computed by the methods used in Bernoulli's theorem. These methods lead to the conclusion that in most of the columns the frequency of B will correspond to its horizontal probability p. It is thus proved that the longer the columns the higher the horizontal probability w_n for a vertical frequency of B corresponding to p. If n goes to infinite values, w_n goes to 1.

We would like to conclude that there is a horizontal probability $w_\infty = 1$ for finding columns for which q equals p. Yet, strangely enough, we cannot infer that there exists even one column for which actually $q = p$ is satisfied, even if this equality is to hold only within certain limits of exactness. That is, the limit of the w_n need not have the meaning of a horizontal probability of columns. This is shown by the construction of the following lattice. To the left of the lattice diagonal line, given by the elements x_{11}, x_{22}, x_{33}, . . . , we put "B" into every place. To the right of the line we continue all rows as random sequences. In this lattice, the conditions for the inference in terms of Bernoulli's theorem are satisfied, since the independence condition (16) holds; and the horizontal probability w_n has the same value as before. But in every column, the vertical probability q is equal to 1, and thus different from p. The reason is that the condition (18) of lattice invariance is not satisfied.

It may be objected that the lattice constructed is a degenerate case, possible only for infinite columns, and that in all practical applications the columns are of finite length. This statement is true; but it would not be permissible to assume that for finite columns we can dispense with the condition of lattice invariance. Using infinite sequences represents an expedient idealization of probability problems; and transferring derived results to finite sequences requires precautions. Our actual lattices are finite in respect to both columns and rows. The concept of limit is replaced for them by that of a practical limit, which merely means early convergence toward the final value of the frequency, that is, the frequency found at the last element. This condition thus requires a rather long stay of the frequency in an interval ϵ of convergence. For a lattice we have similar restrictions on convergence. Let n be the length of a column, and m the length of rows such that among the first m columns a percentage given by w_n shows a frequency of B corresponding to p. We then require that m be small as compared with n; otherwise, the lattice, for all practical purposes, would have the properties of the degenerate lattice described. For this reason, condition (18) of lattice invariance must be taken over, transcribed into a suitable form, into a finitized calculus of probability.

A physical illustration of the degenerate lattice described can be constructed from the diffusion process presented in figure 18, as follows. We write the lattice only for the nitrogen molecules; that is, we have only B-terms in the first column. Assume that the dividing wall is pulled out very slowly through a slit in the bottom of the container. Then the mixing process begins in the upper rows of the lattice (6), whereas the lower rows remain in B for a longer time. This lattice does not satisfy condition (18), and its columns do not reflect the aftereffect probabilities observed along the rows, although the rows are mutually independent and satisfy (16). Even if the columns are of finite length, we thus can distinguish between a normal mixing process and a degenerate mixing process; this distinction is indicated in relation (18).

The attempt to eliminate the condition of lattice invariance exhibits a certain tendency to allot to probabilities in the time ensemble a priority over probabilities in the space ensemble. The tendency springs from the belief that it is preferable to replace a vertical probability by a horizontal probability referring to columns of finite lengths. But it is hard to see on what grounds such an opinion can be justified. If a column segment of finite length is given, which does not satisfy the frequency relations expected, we may say: Go on in the time ensemble; then you will find column segments which satisfy the expected relations. But we may argue as well: Go on along the column, making the segment longer; then you will find the expected relations satisfied. If the latter interpretation of the phrase "it is probable that" is accepted, we have committed ourselves to regarding the condition of lattice invariance as a separate condition which is not translatable into probabilities within the time ensemble. This is easily seen from the degenerate lattice as described, for which the second interpretation would be invalid, whereas the first would be valid. Yet in application to finite sets of observations, probability statements always have the meaning of predictions about future observations; and once this view is adopted, there is no reason to prefer the time ensemble to the space ensemble. A consistent frequency interpretation of probabilities requires a treatment of probability lattices in which limit statements in both the horizontal and the vertical direction are admitted; and within such a theory the condition of lattice invariance is indispensable.

The comparison with the independence condition reveals the logical nature of the condition of lattice invariance. This condition expresses an extended form of independence assumption. In order to derive the inference from time ensemble to space ensemble, it is not enough to know that the rows are mutually independent; they must satisfy the stronger condition of lattice invariance.

The lattice of mixture is the mathematical model of a mixing process. It formulates the mathematical relations the physical counterpart of which is the mixing of molecules, or of other objects that are subject to a stirring mechanism. It shows that a mixing process is governed by three fundamental relationships: aftereffect, independence, and lattice invariance. Aftereffect refers to relations within a row; independence to relations between different rows; and lattice invariance to relations between horizontal and vertical probabilities. We shall see later that these relations can also be satisfied by physical systems which are not agitated by a stirring mechanism, and that the mathematical formulation in lattice language allows us to transfer the derived results to processes which represent the aspect of mixing in a more general sense.

● 13. The Reversibility Objection

We must now investigate whether the statistical solution of the problem of entropy allows us to define a *direction* of time. It may be recalled that the *order* of time was defined in a previous section (§5) by means of causal laws. Taking over this order, that is, assuming that the events of the physical world are ordered in causal chains, we wish to find ways of distinguishing one direction of these chains, to be called positive time, from the opposite direction. That is, we wish to show that time is not only *ordered*, but *unidirectional*. This problem will be investigated in this and the following sections.

A simple objection against the use of a statistical definition of time direction may be answered first. It was mentioned (§7) that, according to the statistical interpretation of the entropy law, the reverse of a thermodynamical process can no longer be called impossible, but can only be said to be very improbable. Now one might argue that contradictory time directions might result, depending on the kind of process we observe: If we happen to encounter one of the improbable processes, it would be assigned a time direction opposite to that of other processes. But this objection is not tenable. Time direction is a property of the causal net as a whole, and is transferred from the net to individual processes. We define positive time as the direction in which most thermodynamical processes occur. If individual processes were to occur in the direction of negative time, we would detect these exceptional cases by the use of the general definition. And we shall also assign a direction to reversible processes, which in themselves do not exhibit directional properties, by coördinating them with the general direction of the causal net.

Furthermore, it may be noted that the statistical version of the definition of time direction does not imply such statements as: It is merely very improbable that the past comes back. We saw above (§5) that a return to the past would constitute a closed causal chain, the exclusion of which is an order property and has nothing to do with time direction. A statistical interpretation of time direction is quite compatible with a physical law that causal chains are never closed. The criticism of this law belongs in another context, in the discussion of the physics of strict laws.

However, difficulties of another kind have arisen for the statistical definition of time direction. These difficulties concern the question whether it is true that changes toward states of higher entropy, or of higher probability, are more probable than changes in the opposite direction. Strangely enough, there exist serious objections against the very argument on which the statistical interpretation of entropy is based—namely, against the principle that the evolution of physical systems tends to go in the direction of higher probability.

The problem which the physicist faces can be formulated as follows. The elementary processes of statistical thermodynamics, the motions and collisions of molecules, are supposed to be controlled by the laws of classical mechanics and are therefore reversible. The macroprocesses are irreversible, as we know. How can this irreversibility of macroprocesses be reconciled with the reversibility of microprocesses? It is this paradox which the physicist has to solve when he wishes to account for the direction of thermodynamical processes and for the direction of time. The statistical interpretation of entropy seems to offer a solution, because it presents the direction of thermodynamical processes as a tendency to proceed from order to disorder. Yet, on closer inspection, we discover that this seeming solution can be questioned, that the reversibility of elementary processes acts as a check on the tendency of proceeding from order to disorder. *This paradox of the statistical direction* must now be studied in detail.

Consider the diffusion process discussed with respect to figure 18, in §12. As before, we put nitrogen into one partition of the container; into the other we put oxygen. The container is completely insulated from its environment. We now take out the separating wall; then the gases will mix. In the thermodynamical interpretation, the entropy thus becomes larger, because the sum of the two entropies of the separate gases can be shown to be smaller than the entropy of the total mixture. In the statistical interpretation, the process goes from an ordered macrostate to an unordered one, and thus from low to high probability; this was the result reached in the discussion in §12.

Now let us perform a fictitious experiment. After the mixture has taken place, we stop each molecule individually and give it a push in exactly the opposite direction, so that it travels back with the same speed as it had when it arrived. This reversal of velocity could be achieved, for instance, by letting each molecule be reflected by a sort of little mirror held squarely in its path; however, all these reflections have to take place at the same time. What would happen now? Obviously, because of the reversibility of elementary processes, the exact reverse of the preceding process would take place; the gases would separate, and, after a while, we would find all the oxygen molecules on one side of the container, and the nitrogen molecules on the other side.

We would like to argue: But you cannot perform this experiment, which is purely fictitious. You cannot artificially construct the arrangement given by reversed molecules; such an arrangement would be extremely improbable, because it would be highly ordered. Hence, the separation process can practically never occur.

It is here that the real difficulty begins. In Boltzmann's gas theory—and the same is true for the versions developed by Maxwell and by Gibbs—the probability of a molecule's having a certain velocity is independent of the sign of the velocity; this is seen from (9, 14) for the equilibrium state. The reversed velocity, therefore, has the same probability as the original one. This does not mean that we should expect the molecules all of a sudden to reverse their velocities spontaneously. It means that a state in which the molecules travel in exactly opposite directions with the same speed as in the first state may develop naturally as a product of the continuous exchange of velocities in collisions, and that such a state has in the course of time the same probability of occurrence as the first state. But it then follows that separation processes must be exactly as frequent as mixing processes.

This argument is called the *reversibility objection*. It was constructed in the early days of the kinetic theory by Boltzmann's colleague in Vienna, J. Loschmidt (1876),[1] and since then has frequently been discussed. Boltzmann has given his own answer, which will be investigated presently.

---

[1] J. Loschmidt, "Über den Zustand des Wärmegleichgewichtes eines Systems von Körpern mit Rücksicht auf die Schwerkraft", *Sitzungsber. d. Kaiserl. Akad. d. Wiss. in Wien*, math.-naturwiss. Cl., Bd. 73 (1876), 2. Abt., p. 139.

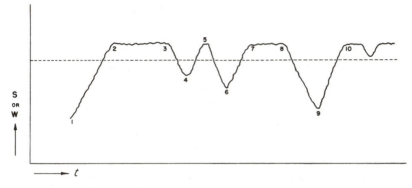

Fig. 19. The entropy curve of a closed system.

Let us first present the objection in a somewhat different version. The changes of the entropy S, or of the probability W, in a closed system which begins to develop in an ordered state, like the gas in the container mentioned, can be represented by a curve as given in figure 19. The initial macrostate A, which we assume to be at place 1, has a low entropy. From this value the entropy rises to a high value at place 2 and then remains near the higher value for a long time. However, it will not keep this value precisely; there will be fluctuations, making the curve go up and down in an irregular fashion. Large fluctuations are much more improbable than smaller ones; but their probability is above zero, and therefore they must occur occasionally. This is an essential feature of probability sequences: every combination of attributes that has a nonzero probability must occur with that nonzero frequency. When we throw a die often enough, a run of a thousand throws showing face 6 on top must eventually occur, because the probability of such a run, though very low, is larger than zero. When we shuffle a deck of cards just after putting all the red cards on top of the black ones, we shall transform this ordered state into a mixture; but if we shuffle long enough, we must by pure chance eventually come back to the original state, because the probability of arriving at such an arrangement is larger than zero. From probability considerations of such a kind we can infer that the gas system must sometimes go through large fluctuations, so large that even a separation of the gases will take place.[2]

---

[2]The same result was derived by Ernst Zermelo and Henri Poincaré, who showed that a mechanical system controlled by Hamilton's equations (10, 1) must eventually return to a state very close to its original state—that is, that the ergodic path must be quasi-periodic. See E. Zermelo, "Über einen

The probability of such an occurrence, of course, is extremely low, and waiting a million years would not give us much chance to see a process of this kind. To quote Boltzmann: "One must not assume . . . that two diffusing gases mix and separate several times within a few days. The time within which we can expect an observable separation is so reassuringly large that any possibility of observing such a process is excluded."[3] However, small fluctuations do occasionally occur. The Brownian motion of small particles under the microscope is the product of such chance fluctuations; it results because the number of molecules hitting the particle from one side is not exactly equal to the number of molecules hitting it from the other side, and may thus be regarded as an indication of fluctuations in the gas pressure for very small areas.

The principles of probability assure us that most of the time the curve will stay at high values. But that is all we can say, and occasional drops to low values cannot be excluded; more than that, they must occur according to these very principles. Looking at the diagram (fig. 19) we see now that every upgrade except the initial one is preceded by a corresponding downgrade. Since in the long run the initial upgrade has a vanishing influence on the relative frequency of grades, we infer that for a curve of infinite length there are as many downgrades as upgrades. This means that transitions to lower entropy are as frequent as those to higher entropy; and this is the statement of the reversibility objection.

The reversibility objection can be raised whether we use the deterministic or the indeterministic interpretation of the evolution of closed systems. The reason is that the objection follows from the very nature of probabilities, and is not linked to the assumption of causal laws behind the molecular statistics. It is true that in the original version, which assumes reversal of the molecular velocities, the objection makes use of the strictness of the laws of mechanics and the reversibility of mechanical processes. But if it is agreed that we can assign to every molecular arrangement of a closed system a definite probability for its occurrence in the course of time, then a curve like that

Satz der Dynamik und die mechanische Wärmetheorie", *Ann. d. Physik u. Chemie*, N.F., ed. by G. and E. Wiedemann, Bd. 57 (1896), pp. 485–494; and Henri Poincaré, "Sur le problème des trois corps et les équations de la dynamique", *Acta Math.*, Tome 13 (1890), pp. 1–270, esp. pp. 67–72. This result has been called the *periodicity objection*, and has an import similar to that of the reversibility objection.

[3]Translated from Ludwig Boltzmann and J. Nabl, "Kinetische Theorie der Materie", in *Encyklopädie der mathematischen Wissenschaften*, Bd. V, Teil 1, Art. 8 (Leipzig, 1907), p. 521.

of figure 19 must result, because each probability value must express itself in a definite frequency of occurrence.

To investigate this result in more detail, let us examine the reversal of molecular velocities for a system that is not completely closed. Let a be the initial microstate in which the gases are separated; and b, the ensuing mixed microstate. By b^* we understand the state resulting when all velocities in b are reversed, from which state the gases return to the state a^* which corresponds to a except that all molecular velocities are reversed. The states a and a^*, which belong to the same macrostate A, are ordered states; since their order is easily visible, we speak here of *explicit order*. The states b and b^* are unordered states. They are states of mixture which are macroscopically indistinguishable and belong to the same macrostate B. But since b^* is of the exceptional kind leading to the ordered state a^*, we may say that b^* contains an *implicit order*. When we now assume that irregular little impacts from the outside occur, we find an important difference between the two kinds of order. Explicit order is *stable;* that is, little impacts will not disturb this order. But implicit order is *unstable;* little impacts will disturb it, in the sense of leading to complete disorder. For instance, when in the process of reversal a few molecules are omitted, the implicit order is destroyed, and the system will not develop into an ordered state. Little changes of this kind in a, however, would not destroy the explicit order, and the system would, as before, develop into an unordered state.

One would like to use this distinction to answer the reversibility objection. All that can be proved in this way, however, is that the reverse process is made even more improbable, not that it is excluded. Irregular impacts may occur in all forms and combinations; and there may be combinations which do not destroy an implicit order. The probability of such combinations of chance impacts is larger than zero, and therefore they must occasionally occur. The disturbances from the outside merely act as a stirring mechanism; they serve to shuffle the sequence of gas states, but cannot prevent it from progressing occasionally in the direction toward order.

These results may now be applied to the discussion of the evolution of the universe. If the universe is spatially finite, we can speak of its state at a certain time t. Such a state is sometimes called a temporal *cross section* through the four-dimensional space-time world; it is defined only after a definition of simultaneity has been given (see fig. 8, in §5). We may assume that the state is given to the exactness of a microstate; but since every microstate determines its corresponding macrostate (see §9), we can speak of the macroprobability, or of the entropy, of an *instantaneous state* of the universe. For a spatially finite universe, therefore, the curve of figure 19 represents not only the development of an isolated gas system, but also the evolution of the universe. Only if the universe is spatially infinite can the probability of its states not be defined, and then we cannot speak of the entropy of the universe. But let us postpone the discussion of this possibility and restrict our investigation, for the time being, to a universe for which a total entropy is definable. However, let us assume that time is infinite, that is, that the causal net is open (see §5).

For such a universe, the curve of figure 19 is to be continued to the left, because time is infinite in the negative direction, too. This means we would have to draw on the left a section going upward from point 1 to higher entropy values in the direction of negative time. The total part on the left of point 1 has thus a shape similar to that of the part on the right, although, of course, it is not the exact mirror image of the part on the right, the curve being irregular in shape. Obviously, however, upgrades and downgrades occur equally often on the curve; the curve, therefore, does not possess a direction.

Now we can ask two kinds of questions. The first concerns an *inference from time to entropy*; the second, an *inference from entropy to time*. Let us begin by studying the first kind of question.

A question of the first kind may have the form: Given the time direction and an initial nonequilibrium macrostate A of entropy S_A; will the entropy S_B of a macrostate B be higher if B is later than A by the time difference Δt? The answer is: If B follows A by a time difference Δt which is not too small, it is more probable that $S_B > S_A$ than that $S_B < S_A$. We derive this result from the curve of figure 19, as follows. Assume that A is situated on the section 5–6 at the intersection with the dotted line of constant height; then a somewhat later state has a lower entropy. But if we select a larger Δt, state B will be in the upper part of section 6–7, and thus have a higher entropy than A. The same result will hold for larger Δt, because B is then situated on the horizontal section between 7 and 8. Only if Δt is still larger will B go to lower entropy values on section 8–9. But this will be an exception, because the horizontal distances between the caved-in parts of the curve are irregular. The probability considered refers to the class of points A lying at the intersections of the dotted line with the curve; and if the same Δt is used for all these points, B will be situated at a higher level in the majority of cases. This result holds whether we select a smaller or a larger Δt, provided it is not too small, and provided we keep the distance Δt between A and B constant while A slides along the horizontal line. This is the meaning of the statement that if A has a low entropy, $S_B > S_A$ is more probable than $S_B < S_A$.

Strangely enough, the same conclusion can be drawn if B precedes A in time. If A slides along the dotted horizontal line while B is to the left of A by the time difference Δt, we find the same result as before: for a constant Δt, it happens much more often that B occupies higher entropy values than lower ones. We find: If B precedes A by a time difference Δt which is not too small, it is more probable that $S_B > S_A$ than that $S_B < S_A$. And we see: The inference from time to entropy leads to the same result whether it is referred to following or to preceding events. This fact expresses the symmetry of time direction for the entropy curve.

These results were derived by Boltzmann and later analyzed in detail by P. and T. Ehrenfest.[4] Boltzmann regarded this consideration as an answer to the reversibility objection.[5] Since it entitles us to expect that an ordered state A will develop into an unordered state B, it is an answer, indeed, when we consider the inference from time to entropy. For instance, when we take out the partition in the container presented in figure 18, we may expect, with high probability, that, when we investigate the gases at some later time, we shall find them well mixed. For a container of ordinary size, a time interval Δt of even a few minutes will lead to an overwhelmingly high probability in favor of an increase of entropy. The probability of the reverse process does not diminish this probability noticeably. Since, however, the same conclusion can be drawn for a state B which temporally precedes A, we see that the inference from time to entropy does not presuppose a time direction of the entropy curve; it merely expresses the fact that the curve possesses upper and lower levels and stays in upper levels much longer than in

---

[4]P. and T. Ehrenfest, "Begriffliche Grundlagen der statistischen Affassung in der Mechanik", in *Encyklopädie der mathematischen Wissenschaften*, Bd. IV, Teil 4, *D* (Leipzig, 1911), Art. 32, p. 43.

[5]See, for instance, Boltzmann's treatment of the reversibility objection in a letter to the editor of *Nature*, Vol. 51 (1895), pp. 413–415. More detailed expositions of this view are found in Boltzmann's first answer to J. Loschmidt—"Bemerkungen über einige Probleme der mechanischen Wärmetheorie", *Sitzungsber. d. Kaiserl. Akad. d. Wiss. in Wien*, math.-naturwiss. Cl., Bd. 75 (1877), 2. Abt., pp. 62–100 (see esp. p. 72)—and in his last publication on the subject matter of gas theory, written in collaboration with J. Nabl, "Kinetische Theorie der Materie", in *Encyklopädie der mathematischen Wissenschaften*, Bd. V, Teil 1, Art. 8 (1907), No. 14, pp. 519–522. In these publications, Boltzmann says repeatedly that "the second law of thermodynamics can never be proved mathematically by means of the equations of motion alone" (see his letter to *Nature*, p. 414). This remark must not be interpreted as meaning that Boltzmann believed a proof like the one given by John von Neumann and George Birkhoff would be impossible. Boltzmann wanted to say merely that no proof is possible according to which the entropy must *always* increase. We can prove only that such increase is very probable.

lower ones. A curve of this kind could be used to define the difference between upper and lower levels, but not to define a difference between left and right, that is, between positive and negative time.

The inferences discussed may be illustrated by the following model. Imagine a plateau of an average altitude of 4,000 feet, which is traversed here and there by canyons of different depths extending from north to south, in such a way that a vertical cross section through this country in a west—east direction resembles the curve of figure 19. A man walks due east along the plateau; when he comes to a canyon he walks down, then up on the other side with the same speed, always keeping his general direction toward the east. Now we can make the following statement. If the man is at an altitude of 2,000 feet (in a canyon), then in two hours he will probably be at a higher level. But we can also say: If the man is at an altitude of 2,000 feet, then two hours before he was probably at a higher level. Both statements are true. This symmetry expresses the fact that a cross section through this country does not define a direction, but does make higher levels more probable than lower ones.

We might suppose that according to these results we must most frequently find the system close to a saddle-point. This conclusion is not correct. It would require that A is both followed and preceded by a state of higher entropy within a small interval Δt. But for small Δt the probability of either of these occurrences is not very high, and the probability of their joint occurrence is lower than that of one occurrence taken alone. It can be shown to be even lower than the product of the two; that is to say, the occurrences are not independent, but somewhat contraindicative. Hence, there is only a small probability that we are close to a saddle-point.

Let us now examine a question of the second kind, which concerns an inference from entropy to time. It has the following form: Given two states A and B such that the entropy S_B of B is larger than the entropy S_A of A; which of the two states is the later one?

From the preceding result it follows that this question cannot be answered by the use of the entropy curve. That A precedes B is just as probable as that B precedes A. This conclusion also follows from our discussion of the reversibility objection (see pp. 111–113). Now it is obvious that if we want to define a direction of time, we must be able to answer questions of the second kind, concerning an inference from entropy to time. That we can answer questions of the first kind, concerning an inference from time to entropy, does not help us when we do not yet know the time direction. And we see that the reversibility objection remains valid when we wish to use the entropy curve for a definition of time direction, because the curve does not supply such a direction. The paradox of the statistical direction remains unsolved: the reversibility of the elementary processes is transferred to macro-processes; therefore, the statistical direction of isolated systems, just like the direction of the elementary processes, cannot be proved to be unique.

If we wish to solve the paradox and to answer questions of the second kind, we must therefore look for other methods. The entropy curve, and with it, the reversibility objection, pertain to a time en-semble, namely, to the history of one system. We shall discover better methods of answering our question when we abandon the time ensemble and turn to a study of a space ensemble.

● *14. The Time Direction of the Space Ensemble*

If we wish to find a way of defining a direction of time, it is advisable to study the actual procedure which is used when inferences concern-ing time direction are made.

We said that, if $S_B > S_A$, the theory of the time ensemble tells us that A might be earlier as well as later than B. In all practical applications, however, we shall be very willing to infer that, if $S_B > S_A$, A is earlier than B. For example, if one observer tells us he saw the gases in a container rather well separated, though they were not divided by a par-tition, and another observer informs us that he saw the gases well mixed, we shall conclude that the second observation was made later than the first. We shall add the further conclusion that originally the gases were separated by a partition, and that someone must have re-moved the partition shortly before the first observation was made. This means that, rather than proceeding on the assumption that the gas system was closed all the time, we assume that it was originally in interaction with its environment; and we conclude that the improbable state is the product of this interaction rather than the result of a separation process produced by mere chance in the history of the closed system. Since this inference includes a reference to systems other than the observed one, it is not covered by the curve of figure 19 and must therefore be more closely examined.

Our environment is rich in processes which, either as a natural product or through the intervention of man, create as parts of their results ordered subsystems, which from then on remain isolated and run through an evolution toward disorder, as shown in the initial sec-tion 1–2 of the curve of figure 19. By "isolated" we do not mean complete isolation; it is sufficient if the process within the subsystem represents energy exchanges which are large compared with the interaction with the environment. The refrigerator was mentioned in §7 as an illustration of subsystems of low entropy. If we take an ice cube out of the refrigerator and put it into a glass of water, we have an isolated system, though the thermic isolation is none too good; the melting of the cube is represented in section 1-2 of the entropy curve of figure 19. A rock embedded in snow is heated by the sun; during this process, the system consisting of the sun and the rock increases its entropy because of the absorption of radiation by the rock. At night, the rock and snow together form a system which, because of its inner temperature differences, has a low entropy; and the cooling of the rock and melting of the snow represent a transition to higher entropy. Chemical processes such as fire or the breathing of living organisms create temperature differences in their environment which lead to compensation processes going from order to disorder. The creation of ordered initial states—of ice cubes floating in a glass of water, of temperature differences resulting from chemical processes—is not achieved in these cases in isolated systems which undergo an entropy decrease, but in subsystems of comprehensive systems the total entropy of which goes up while the subsystem is put into a state of relatively low entropy. Nature abounds in *branch systems* of this kind, that is, systems that branch off from a comprehensive system and remain isolated from then on for some length of time. Their evolution begins with an ordered state, that is, a state of relatively low entropy, and progresses toward disorder, that is, toward relatively high entropy. We use the word "relatively" here in order to indicate that this entropy is referred to the subsystem, not to the universe or the main system.

These are the observational facts to which the statistical definition of time direction must be referred. When we infer from the inequality $S_B > S_A$ that in all probability A was earlier than B, this probability is of a kind different from the one discussed in connection with the reversibility objection: it refers, not to the sequence of states of the isolated system to which the states A and B belong, but to the series of similar systems, conceived as an ensemble of branch systems. That is, it refers, not to a time ensemble, but to a space ensemble.

In order to study probabilities of this kind, let us first introduce a certain simplification. Let us consider a long upgrade of the entropy curve of the universe and restrict our statistics to systems branching off from this upgrade, assuming that these branch systems remain isolated for an infinite time. Figure 20 may indicate this schematization of the problem. We may assume that the curves of the branch

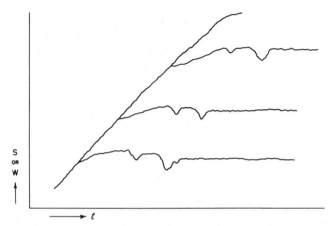

Fig. 20. The initial upgrade of the entropy curve of the universe. Some isolated systems branch off, remaining isolated for an infinite time. (Simplified schematization.)

systems are drawn to a larger scale than the total entropy curve of the universe.

We will now symbolize the world lines of the branch systems by a probability lattice, as introduced in (12,6). In contrast to the lattice of molecules considered there, the present lattice refers to systems; that is, each row is the world line of a whole system, and we write down the transition of the system from one state to the next one, and so on, each state being characterized by its entropy. In order to simplify the investigation, we will avoid reference to continuous probability sequences by cutting the history of the system into discrete events, which follow one another at short time intervals Δt, in the same way as was done for the histories of molecules. Denoting an individual event in the history of the system by "y", we arrive at the probability lattice:

$$
\begin{aligned}
& y_{11}\ y_{12}\ y_{13}\ y_{14} \cdots y_{1i} \cdots \\
& y_{21}\ y_{22}\ y_{23}\ y_{24} \cdots y_{2i} \cdots \\
& \cdots\cdots\cdots\cdots\cdots\cdots\cdots\cdots \\
& y_{k1}\ y_{k2}\ y_{k3}\ y_{k4} \cdots y_{ki} \cdots \\
& \cdots\cdots\cdots\cdots\cdots\cdots\cdots\cdots
\end{aligned}
\tag{1}
$$

The attribute space of each row is given by the energy surface of the phase space of the branch system. If A is a certain value of the entropy, defined within a small interval of exactness, then A represents a certain area in the attribute space; and if the state y_{ki} of the system has this entropy value, we say that the element y_{ki} is a member of the class A. The first element y_{k1} of a row is in a highly ordered state; in the succeeding elements the order gradually disappears until a state of disorder is reached. That is, each row has an aftereffect and changes its attributes gradually. If the rows are continued for an extremely long time, reversals toward lower entropy would occur, as indicated in figure 20.

We thus have here a lattice of mixture and can apply the notation and the theorems developed in §12. If A and B are states of different entropies, there exists a horizontal probability that A is followed by B after m elements; this probability is written:

$$P\left(A^{ki}, B^{k,i+m}\right)^i . \tag{2}$$

If A is a state of low entropy and B is one of high entropy, and if m is not too small, the probability (2) is very high.

We can also ask for the probability that A is preceded by B at a distance of m elements. This probability results from expression (2) upon reversing the sign of m, assuming that $m < i$. Because of the directional symmetry of the row, discussed with reference to the reversibility objection, we have here the relation

$$P\left(A^{ki}, B^{k,i+m}\right)^i = P\left(A^{ki}, B^{k,i-m}\right)^i . \tag{3}$$

It is this equality which prevents us from using a row for the definition of time direction.

We can also symbolize a vertical probability, in which we count the frequency along a column:

$$P\left(A^{ki}, B^{k,i+m}\right)^k . \tag{4}$$

This expression means, once more, the probability that A is followed by the state B after m elements in the same row; but here this probability is counted in the vertical direction, that is, in an ensemble of different branch systems. If A and B have the same meaning as in (2), we have the relation

$$P\left(A^{ki}, B^{k,i+m}\right)^k = P\left(A^{ki}, B^{k,i+m}\right)^i , \qquad m > 0 . \tag{5}$$

This means that we assume the lattice to satisfy the condition of lattice invariance, introduced in (12, 18). It is an empirical assumption confirmed by many experiences.

We see here the physical significance of the condition of lattice invariance for the space ensemble of branch systems. It is difficult to come to reasonable estimates concerning the probabilities in the space ensemble. But we know the probability metric in the time ensemble very well, because this metric is determined by Liouville's theorem (see §10). The assumption of lattice invariance allows us to transfer this metric from the time ensemble to the space ensemble.

However, lattice invariance does *not* apply to probabilities concerning *preceding* events, because it was assumed that all branch systems begin with states of low entropy. *Equation* (5) *is false if* $m < 0$. If, again, A is a state of low entropy and B a state of high entropy, we find, counting vertically, that in most cases the entropy increases toward the right, as long as we are not too far away from the beginning of the rows. Consequently we have for small i the inequality

$$P\left(A^{ki}, B^{k,i+m}\right)^k > P\left(A^{ki}, B^{k,i-m}\right)^k , \qquad 0 < m < i . \tag{6}$$

In contrast to (3), *we find here: the probability that a low-entropy state is followed by a state of high entropy is greater than the probability that the same low-entropy state is preceded by a state of high entropy.* The reason that the reversibility objection does *not* apply here is that we are dealing with vertical probabilities, that is, with statistics of an ensemble of branch systems, and not with the evolution of just one system. Let us call a horizontal probability a *one-system probability*, and a vertical probability a *many-system probability*. The solution of the paradox raised by the reversibility objection is then based on the distinction between one-system and many-system probabilities; or, what is the same, on the distinction between probabilities referring to a time ensemble and probabilities referring to a space ensemble.

If all the branch systems considered in the lattice are of the same physical kind and their initial states A are the same, and if A is again a state of low entropy, we have, using (5):

$$P\left(A^{k,m+1}, B^{k,m+i}\right)^m = P\left(A^{k1}, B^{ki}\right)^k = P\left(B^{ki}\right)^k . \tag{7}$$

This means that the absolute vertical probability of finding B in the i-th column (last term) measures directly the relative horizontal probability of finding the state B following a highly ordered state A after $i - 1$ steps (first term). The high one-system probability that a state B of high entropy follows a state A of low entropy, is here reflected in the many-system probability of B.

If the entropy of B has a medium value, intermediary between a low and a high entropy, the state B will not be very frequent in the history of one system. Thus we have

$$P\left(A^{k,m+1}, B^{k,m+i}\right)^m > P\left(B^{k,m+i}\right)^m = P\left(B^{ki}\right)^i . \tag{8}$$

However, such a state B of an intermediary entropy value can be rather frequent in the ensemble of branch systems; this is stated in (7), which relation, in combination with (8), leads to the result

$$P\left(B^{ki}\right)^k > P\left(B^{ki}\right)^i \; . \tag{9}$$

This inequality, which corresponds to relation (12, 9), expresses the influence of the highly ordered initial state A in the beginning of each row. But this influence dies down, and we have

$$\lim_{i \to \infty} P\left(B^{ki}\right)^k = P\left(B^{ki}\right)^i \; . \tag{10}$$

This relation corresponds to relation (12, 10). Furthermore, the probability on the right in (6) increases and converges to the value of the probability on the left, the latter remaining constant in this process. For the limit $i \to \infty$, therefore, relation (6) turns into an equality corresponding to (3).

This lattice represents a lattice of mixture in which all elements of the first column are in the same state, described by relations (12, 9-14). A lattice of this kind defines a direction through the frequency ratios in its initial columns. And the space ensemble defines a time direction perpendicular to this direction if it is embedded in a lattice of mixture.

These considerations show that the lattice of systems can be regarded as a mixing process in the same sense as the lattice of molecules. This is true in spite of an important physical difference between the two examples: molecules collide, and their interaction serves as a shuffling mechanism, whereas the systems do not interact with one another and are not shuffled by a physical mechanism. Yet the function of such a mechanism is taken over by the laws of chance governing independent chains of events. These laws, which are expressed in the conditions of independence and lattice invariance, combine to destroy gradually the order expressed in the arrangement of the first column, which contains only low-entropy states, namely, those states in which the systems branch off from the main system.

When we understand by a mixing process a lattice controlled by the mathematical relations developed in §12, we may say that the time direction supplied by the ensemble of branch systems originates from a mixing of systems, in the same sense that the time direction of a diffusion process results from a mixing of molecules. Both kinds of ensembles furnish a mixing process, because they satisfy the conditions of a lattice of mixture, in which the first column consists of selected states. It appears that mixing processes, in the most general sense of the term, are the instruments which indicate a direction of time. They can do so because they translate the symmetry of horizontal probabilities into an asymmetry of vertical probabilities, or the directional symmetry of the time ensemble into an asymmetry of the space ensemble. This leads to two distinct meanings of the phrase "probability that a low-entropy state is preceded by a state of high entropy", and thus makes it possible to account for the inference from entropy to time.

A remark for the logician may be added. In the investigations presented in this book, the term "probability" is always assumed to mean the limit of a relative frequency, as explained in (12,2). This frequency interpretation of probability, now generally accepted among scientists, can account for all the uses of the term "probable" that are verifiable in terms of predictions and therefore have predictive value. This means that if a probability statement is true in this interpretation, its use for predictions can be shown to be advisable. A complete mathematical and logical theory of probability based on the frequency interpretation has been constructed.

However, if we ask for the probability p that a given state A of low entropy is preceded by a state of high entropy, the question refers to a single case. It follows from the frequency interpretation that a probability statement concerning a single case has no meaning. In order to justify the application of probability statements to a single case, these statements have to be replaced by statements referring to frequencies. This requires, first, the selection of a suitable reference class;[1] but, secondly, we also have to select a suitable sequence.[2] The example of the probability p illustrates the need for the second requirement: the reference class A and the attribute class B do not determine this probability; using the same classes we obtain a high, or a low, value for p according as we select as our sequence a row or a column, respectively. This ambiguity, which is the necessary consequence of lattice arrangements of this kind, shows that it is impossible to speak of the probability of a single case in a meaningful way. Logicians who reject the frequency interpretation and attempt to define the probability of a single case as having a specific meaning not referring to frequencies, should test their theories on problems like the reversibility objection; they would then soon discover that their systems do not cover the use of probabilities in practical applications.

The vertical probabilities occurring in (6) are not yet identical with the probabilities used in the inference from entropy to time, because the probabilities in (6) presuppose a knowledge of time direction. However, a probability v going from entropy to time can be constructed[3] from the probabilities used in (6). In order to show that the probability v also refers to columns, and not to rows of the lattice, I will give here the exact symbolization of this probability.

---

[1] See *ThP*, p. 374.

[2] See *ibid.*, p. 376.

[3] [This inference seems to presuppose the definition in §15 of the present study (p. 127).—M. R.]

We first define a probability v_{im} referring to the i-th column as follows. The reference class is given by those states y_{ki} which belong to A *and* which are followed *or* preceded in the m-th element by a state B. The attribute class is given only by the cases of the first kind, that is, states y_{ki} belonging in A and *followed* in the m-th element by a state B. In other words, out of all the cases in which a state A is followed or preceded at the m-th place by a state B, we count only the first kind of cases. This probability v_{im} is symbolized in the following form:

$$v_{im} = P\left(A^{ki} . B^{k,i+m} \vee A^{ki} . B^{k,i-m}, A^{ki} . B^{k,i+m}\right)^k . \qquad (11)$$

The hook sign means "or"; the period sign means "and". A probability of this kind, in which the reference class has the form of a disjunction, is solved by means of the rule of reduction,[4] as follows:

$$v_{im} = \frac{P(A^{ki} . B^{k,i+m})^k}{P(A^{ki} . B^{k,i+m})^k + P(A^{ki} . B^{k,i-m})^k - P(A^{ki} . B^{k,i+m} . B^{k,i-m})^k} . \quad (12)$$

Each of these terms has the general form $P(A.B)$ and can be written, according to the general multiplication theorem, as $P(A) \cdot P(A, B)$. The first factor $P(A^{ki})^k$ thus resulting is the same for all terms and drops out (it is greater than 0, because we assume that the state A does occur in the column i). We find

$$v_{im} = \frac{P(A^{ki}, B^{k,i+m})^k}{P(A^{ki}, B^{k,i+m})^k + P(A^{ki}, B^{k,i-m})^k - P(A^{ki}, B^{k,i+m} . B^{k,i-m})^k} . \quad (13)$$

Now for all practical purposes the probability on the right side of (6) will be extremely small if $S_B > S_A$ and i is not too high. Therefore, the second and third terms in the denominator drop out, since it is easily shown that the third term is even smaller than the second. Hence, we find

$$v_{im} \sim 1 , \qquad (14)$$

where the curl sign means "is approximately equal to". The probability v, in which i and m are not specified, results as the average, or mean value,[5] of the various v_{im}, and we have

$$v = M_I\left(v_{im}\right)_{im} \sim 1 . \qquad (15)$$

---

[4] *ThP*, p. 100, formula (17). In the symbolization given above, I omit the general reference class and use the absolute notation introduced in *ThP*, p. 106.

[5] For the notation, see *ThP*, p. 180.

The subscript "I" added to the symbol of the mean may indicate that the values i must not be too high; that is, like the inequality (6), relation (15) applies only to the initial columns of the lattice.

From the discussion of the lattice (1) earlier in this section, the result (15) appears obvious; but the symbolic derivation is helpful because it discloses the logical structure of the inferences used. Inferences of this kind are employed in the many situations in which we infer that the ordered state precedes the unordered one. For instance, if one aerial photograph taken during the war shows a city intact, whereas another one shows its houses bombed into ruins, we know which one was taken first.

Another kind of inference goes toward the past from a given state B which has a medium entropy value and concludes that B was preceded by a state A of lower entropy. Such an inference is based directly on the relation

$$P\left(B^{ki}, A^{k, i+m}\right)^k \sim 1 \ , \tag{16}$$

which holds for the initial columns of the lattice (1) if $S_B > S_A$. For instance, when we find a burning cigarette on an ash tray, we infer that, shortly before, it was not burning and had its full length; or if in the evening we find a rock that is warmer than the surrounding air, we infer that it was still warmer during the daytime and was heated by the sun. Note that when we say the entropy of the earlier state was lower, we refer to the total system including all changes that have occurred. For instance, the radiation emitted by the cooling stone is then to be included. Relation (16) shows that many-system probabilities allow us to infer that in the past the entropy was lower; we have seen that such a conclusion cannot be drawn for a one-system probabilities. Only the theory of the *space* ensemble gives the answer to the reversibility objection.

We have carried out these computations on the assumption that *all* rows of the schema (1) begin with the states of relatively low entropy. If this is not quite true, but merely the vast majority of rows are of this kind, the same results can be derived. We need not give the demonstration for this slightly more general case, because the probability relations remain unchanged.

• 15. *The Sectional Nature of Time Direction*

The solution of the reversibility objection by the introduction of probabilities referring to branch systems indicates the way of defining a direction of time. However, our discussion has thus far been restricted to the schematized arrangement of branch systems indicated in figure 20. This schema is inadequate in two ways. On the one hand, it is restricted to one upgrade of the entropy curve; on the other hand, it represents branch systems as infinite in time.

First, we know that continuing the main curve beyond the upgrade represented in figure 20 leads to a horizontal section and then to downgrades and upgrades alternating in the irregular fashion indicated in figure 19. Secondly, we know that branch systems are usually isolated only for a limited time and then are reconnected to their environment. If we put coffee and cream into a thermos bottle, the resulting mixture will usually not stay in the bottle for too long a time; if the rock heated by the sun melts the snow, the water seeps into the ground and enters into the large system of the surrounding soil.

Combining the first consideration with the second, we arrive at a curve like that indicated in figure 21; however, only one upgrade and one downgrade are drawn in this diagram, as we may assume the curve to be continued on both sides in a similar way. A few branch systems, returning to the main curve, are indicated on each grade.

When we wish to transfer our previous probabilities to this example, difficulties arise. That the horizontal sequences are here not infinitely long need not disturb us, because for all practical applications sequences of a finite length are sufficient if they are long enough. The columns in schema (14, 1) are finite, too, because the number of branch systems for an upgrade of the main sequence is finite; but, as has already been pointed out, this number is extremely high. However, in the arrangement presented in figure 21 the sequences given by the branch systems are not long enough to afford realizations of the horizontal probabilities; they contain only the initial upgrades of the branches drawn in figure 20 followed by a section of the horizontal part, and then return to the main curve. In branch systems actually observed by us, a separation of their constituents never occurs. Only for the main curve can the horizontal probabilities be realized, since this curve is of an infinite length.

In contrast, the vertical probabilities are realized in terms of frequencies; but here the conditions of schema (14, 1) are not satisfied. Only for the upgrade of the curve in figure 21 do the branch systems begin with a state of low entropy, as indicated by the points 1, 3, and 5 of the diagram. When we come to the downgrade, always proceeding in the same direction, the branches begin at states of high entropy, namely, at the points 7, 9, and 11; and they end at points of low entropy, namely, at 8, 10, and 12, respectively. When we now continue

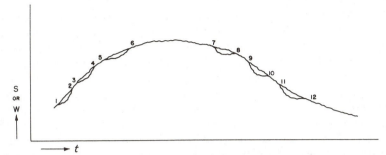

Fig. 21. An upgrade and a downgrade of the entropy curve of the universe. Some isolated systems branch off and return to the main system.

the lattice (14, 1) by using the branch systems of the downgrade and of further sections of the main curve, we find transitions to high and to low entropy in equal frequencies, and relation (14, 6) is replaced by an equality of the two probabilities. Thus the definition of a time direction breaks down.

It follows that a time direction can be defined only for sections of the total entropy curve. Here the inequality (14, 6) applies and allows us to give the definition:

DEFINITION. The direction in which most thermodynamical processes in isolated systems occur is the direction of positive time.

Assume, for instance, that our universe is at present on an upgrade of its entropy curve. Statistics accessible to us will then always concern vertical, or many-system, probabilities restricted to this upgrade of the total curve. Horizontal probabilities like those used in (14, 3) cannot be observed by us in the form of frequencies, because the time span of possible observations is not long enough. Therefore, observable probabilities will always satisfy the inequality (14, 6) and will supply the definition of positive time. We see, however, that these statistics merely express the general trend of the section of the entropy curve on which we happen to live. The total entropy of the world in its present state is not too high: the universe has large reserves in ordered states, so to speak, which it spends in the creation of branch systems and thus applies to provide us with a direction of time.

This consideration leads to peculiar consequences. The time definition carried through for the upgrade of the total curve can be applied equally well to a downgrade. Here, however, it determines the opposite direction as the direction of positive time: judged from the occurrences in this section, the universe travels once more toward higher entropy, if we define positive time correspondingly. It follows that we cannot speak of a direction for time as a whole; only certain sections of time have directions, and these directions are not the same.

The first to have the courage to draw this conclusion was Ludwig Boltzmann.[1] His conception of alternating time directions that are merely sectionally defined by statistical processes represents one of the keenest insights into the problem of time. Philosophers had attempted to derive the properties of time from *reason;* but none of their conceptions compares with this result that a physicist derived from *reasoning* about the implications of mathematical physics. As in so many other points, the superiority of a philosophy based on the results of science has here become manifest. There is no logical necessity for the existence of a unique direction of total time; whether there is only one time direction, or whether time directions alternate, depends on the shape of the entropy curve plotted by the universe.

Boltzmann has made it very clear that the alternation of time directions represents no absurdity. He refers our time direction to that section of the entropy curve on which we are living. If it should happen that "later" the universe, after reaching a high-entropy state and staying in it for a long time, enters into a long downgrade of the entropy curve, then, for this section, time would have the opposite direction: human beings that might live during this section would regard as positive time the transition to higher entropy, and thus their time would flow in a direction opposite to ours. Since these two sections of opposite time directions would be separated by aeons of high-entropy states, in which living organisms cannot exist, it would forever remain unknown to the inhabitants of the second time section that their time direction was different from ours. Similarly, two sections starting from a saddle-point in opposite directions would be separated by long periods in which life could not sustain itself, because in a low-entropy state there would be high temperatures localized in lumps of matter, like glowing gas balls. Life is restricted to the temperate zones of transition in the entropy curve. Thus an alternation of time directions would involve no contradiction to experiences accessible to us. Perhaps we are, indeed, inhabitants of a second section, in which the entropy "really" goes down, without our knowing it.

What do such statements mean? Obviously, it has no meaning to say of one of the sections that its entropy "really" goes up, or that its time direction is "really" positive. Boltzmann compares this problem with the question whether we or our antipodeans "really" have an upright posture. To say, however, that the universe consists of separate time threads pieced together in opposite directions does have a meaning, because time *order* can be defined in classical mechanics (see §5)

---

[1]Ludwig Boltzmann, *Vorlesungen über Gastheorie*, Bd. II (Leipzig, 1898), pp. 257–258.

and does not presuppose entropy. We may therefore speak of a *super-time* which orders the curve even in sections of equilibrium, where the entropy remains practically constant, or at saddle-points, where the entropy gradient reverses its direction. Supertime has no direction, only an order, whereas it contains individual sections that have a direction, though these directions alternate from section to section. For the horizontal parts of the entropy curve, however, a time direction cannot be meaningfully defined.[2]

How do we know that our universe is at present on a long grade, which we call an upgrade? This conclusion cannot be based merely on the fact that at present the entropy of the universe is not too high. Using only one-system probabilities, we might infer that in the future the entropy will be higher; but we saw that the same inference could be made toward the past, and we might therefore just as well conclude that in the past the entropy was higher. Even both conclusions could be true, and we might be at a saddle-point of the curve. When we insist that the entropy was lower in the past, we use other methods of inference.

There are two ways of making such an inference. First, we can use causal laws, that is, the deterministic method; second, we can use probabilities in branch systems. Let us first study the deterministic method.

We saw that the laws of mechanics define an order of time and also allow us to infer from a given state of a system its previous and its future states. The laws of planetary motion can be used not only to foretell the future state of the planetary system, but also to infer the states of this system in previous times. The radiation emitted by the sun and traveling through space is controlled by the differential equations of Maxwell and represents a causally ordered process like the motions of the planets. We know that the universe is very sparsely filled with matter; the sun emits its radiation, therefore, into a practically empty space. Now the spreading of a cloud of radiation into a vacuum represents a process of growing entropy; but this is a process which cannot be reversed by chance alone and requires intervention from the outside for its reversal. Only walls of matter reflecting the radiation could reverse its path. The inference concerning the past of our universe can therefore be made in terms of practically deterministic methods and is based on a detailed knowledge of the universe in its

---

[2]Boltzmann also speaks (*ibid.*, p. 258) of the possibility that the universe may consist of different domains, separated by wide empty spaces, which simultaneously have opposite time directions. See the discussion in §16 of the present study, especially pp. 139–140.

present state. The mere knowledge of its present entropy value would not suffice to permit the inference. We may recall the deterministic interpretation of the ergodic path (§10); we saw that if the location of the phase point is known, its past and future path is calculable. By similar methods, applied to the macrocosm, we can compute the past and the future of the solar system and even of galaxies.

In application to the entropy curve of the universe, we may interpret these results as stating a strong aftereffect of the probability sequence of its states. It is this aftereffect which produces long upgrades and downgrades of the curve and to which we owe the existence of sectional time directions. But the actual inferences toward the past are not merely based on such general probability considerations; they make use of detailed state descriptions, going much beyond classifications into states of entropy, and they employ causal laws referring to the temporal order of these states.

The nature of this inference may be illustrated by reference to a diffusion process in a gas system. When we observe a system of two gases in a partly mixed, partly separated, state, that is, a state of medium entropy value, and we know that the system has been isolated forever, we infer that in the past the entropy of the system was higher, for the same reason that we infer it will be higher in the future. But if we were able to observe the position and velocity of each molecule separately, we could compute the past of the gas system by the use of the laws of mechanics and would not employ an inference in terms of entropy. In this way, we might find out that the system was previously in a state in which its components were completely separated and therefore its entropy was even lower. A detailed knowledge of present conditions thus makes the entropy inference dispensable and opens the way to the use of causal laws for the acquisition of knowledge about the past. Although we cannot use such methods for gases, we can employ them for ascertaining the past of the universe, because we possess a rather detailed knowledge of its present state and are not restricted to a general inference in terms of entropy.

Let us now discuss the second type of inference directed toward the past of the universe. We use here relation (14, 16), which holds for the initial columns of a lattice of branch systems. Inferences of this kind are made, for instance, in geology, when the crust of the earth is regarded as the product of a cooling process; or in astronomy, when the moon's rotation is regarded as having been slowed down through a sort of inner friction from "tides" so as to adjust itself to the moon's revolution around the earth. Our knowledge of the past is essentially based on the existence of a plurality of branch systems, in each of which the entropy has gone through the initial upgrade of its curve, or is still doing so. It was shown in §14 that, in contrast to inferences based on one-system probabilities, an inference from time to entropy based on many-system probabilities is not symmetrical with respect to past and future. It can lead to lower entropy values for the past.

Once more we see that many-system probabilities are required to account for the inferential methods of the physicist. If the kinetic theory were restricted to one-system probabilities referring to isolated systems, it could not inform us about the past of our universe. And deterministic methods are too limited in their applications to supply a sufficient knowledge of the past. The evolution of species, for instance, one of the greatest discoveries of biology, could not have been inferred by the deterministic methods of the physicist. It is based entirely on probabilities in branch systems, as will be shown in the following section (§16).

Returning to the problem of the entropy of the universe, we conclude, using the methods discussed, that in the time direction which we call the future the entropy of the universe will be higher, and that it was lower in the direction which we call the past; in other words, that we live on a long upgrade of the entropy curve. However, the entropy increase of the total universe as such is not the reason for calling the corresponding time direction positive. In our daily experiences, we are not much concerned about the entropy of the universe. Time direction is expressed for us in the directions of the processes given by the branch systems with which our environment abundantly provides us. All these processes go in the same direction, namely, the direction of growing entropy. We saw that this fact is closely connected with a general entropy increase of the universe; therefore, it is through its reiteration in branch systems that the entropy growth of the universe dictates to us a direction of time. The universal increase of entropy is reflected in the behavior of branch systems, so to speak; and only this reflection of the general trend in many individual manifestations is visible to us and appears to us as the direction of time.

The existence of a long upgrade of entropy, though a *necessary* condition for the phenomenon of time direction, is therefore not a *sufficient* condition. Time direction becomes apparent to us only because the upgrade contains a large number of situations in which subsystems branch off, disclosing in their further development the universal growth of entropy. This growth would not become evident unless the entropy of the subsystem initially had a low value as compared with the entropies of other possible states of the subsystem while it remains isolated. Our universe satisfies this condition. It owes its time direction, therefore, to the branch structure of the upgrade of its entropy curve on which we live.

I should like to believe that considerations of this kind determined Boltzmann's views. He speaks of "small isolated systems" which in the "beginning" are always in an improbable state.[3] But Boltzmann's presentation is too brief in its philosophical parts to admit of an unambiguous interpretation. It seems he did not see that the assumption of a branch structure as described in this section goes beyond the assumption that our universe is at present on a slope of its entropy curve. Boltzmann writes: "The fact that actually a transition from a probable to an improbable state does not occur as often as the opposite one, may be sufficiently explained by the assumption of a very improbable initial state of the universe around us. As a consequence, a given system of bodies entering into interaction is found, in general, in an improbable state.'"[4] This argument can make plausible the existence of a branch structure as described above. But it must be inquired what kind of assumption is hidden behind this plausible inference. This question will be taken up in §16.

Furthermore, although the passage quoted refers obviously to frequencies counted in branch systems, there is in Boltzmann's writings no explicit formulation concerning the probabilities referred to this ensemble. As far as the inference from entropy to time is concerned, the reversibility objection cannot be answered by reference to one-system probabilities, but only by the use of many-system probabilities referring to branch systems. We cannot construct for horizontal probabilities an analogue of the inequality (14, 6) by restricting the counting of the horizontal probability to an upgrade of the curve, because, in an isolated system, an entropy state A occurs only once in one section. When we wish to express a probability like (14, 2) in terms of frequencies, we must therefore count across many sections, and thus arrive at the equality (14, 3). Since a sectional count is not possible for horizontal probabilities, such probabilities cannot define a sectional time direction.[5] The ultimate answer to the reversibility objection cannot be given with reference to one isolated system, but requires reference to a plurality of systems which in their beginnings are not isolated, branching off from the total system.

Let us add a remark about a spatially infinite universe. For an arrangement of an infinite number of molecules, a probability, and

---

[3]Boltzmann, *op. cit.*, pp. 257–258.

[4]*Loc. cit.*

[5]A sectional statistics for one-system probabilities might be attempted as follows. A class Q may be defined as including many states A of entropy values, which, however, are all states of low entropy. A class R may include somewhat higher entropy values, although R may overlap with Q. By the probability $P(Q^{ki}, R^{k, i+m})^i$ we now understand the probability that a state y_{ki} belonging to Q is followed after m elements by a state $y_{k, i+m}$ belonging to R

consequently an entropy, is not definable. One might try to define an average entropy s in the following way: We compute the entropy S of an arrangement for a three-dimensional volume V of space and define a specific entropy s by the relation

$$s = \lim_{V \to \infty} \frac{S}{V} \,. \tag{1}$$

However, we do not know whether this limit exists; and if it does exist, it might be equal to 0, or to some constant, and stay there all the time. It might also vacillate irregularly. We are therefore unable to prove that s increases all the time, or follows the pattern of figure 19.

But it is possible to define a direction of time in another way. A system that is spatially separated from other systems, like a star, or a galaxy, cannot be regarded as isolated, because it loses heat through radiation into space. If we disregard the emitted radiation, the entropy of the system goes down, because the system cools off. Only if we include the cloud of emitted radiation spread through space will the entropy of the system go up. We may therefore attempt to say: The evolution of any system can be shown to represent an increase of entropy if the system is incorporated into a more comprehensive system that is sufficiently large.[6] If this principle is true, a unique direction of time for an infinite universe would be defined, although there is no such thing as the entropy of the universe.

At the present state of cosmology, it is very difficult to come to a conclusion concerning time as a whole. Moreover, in order to account for the time direction of the section of the world that is accessible to our experience, it is not necessary to assume that time has one direction throughout. This is a most important result of Boltzmann's investigations: the question of the direction of time as a whole must be separated from the question of the time direction observable by us. Even if the order structure of time is not the open one usually assumed —that is, even if time as a whole is closed, a conception mentioned

---

and such that its entropy is higher than that of y_{ki}. Such probabilities can be realized by frequencies along one upgrade of the curve. For instance, when cement is slowly mixed with sand, we see the process going on and observe that each partly ordered state is followed by a somewhat less ordered state. But such probabilities are the result of an averaging process; their exact definition requires reference to probabilities like (14, 2), which refer to precise states A, and which are summed up according to the rule of reduction (*ThP*, pp. 97 and 234), assuming an antecedent probability of A (which depends on the way we select our observations within Q). And probabilities like (14, 2) cannot be sectionally defined.

[6]Such a formulation was given by Max Planck, in his *Vorlesungen über die Thermodynamik* (Leipzig, 1911), p. 104.

in §5—there still remains the possibility that sections of it have directions which would be statistically expressed in the same way as for the temporally open universe.

With his statistical definition of a time direction that is restricted to a section of the total time of the universe, Boltzmann has shown the way to solve the paradox of the statistical direction, the problem of reconciling the unidirectional nature of macrotime with the reversibility of microprocesses. This result is Boltzmann's great contribution to physics and to philosophy. And the significance of this result is not diminished by the fact that he based his considerations on a mechanical model of the atom, a model which is no longer adequate. It will be seen, on the contrary, that the quantum physics of our day is in need of Boltzmann's ideas just as much as the physics based on Newton's mechanics, for the very reason that this modern physics, too, did not discover irreversibility in its elementary processes. The statistical nature of time direction appears to be the ultimate outcome of all inquiries into the nature of time. This fact does not detract from the import which the directedness of temporal relations has for our conception of time. On the contrary, it will be seen that experiences in our daily environment lend support to the statistical interpretation of time direction, that they tell us directly, so to speak, what Boltzmann found through abstractions from thermodynamical processes. With his statistical analysis Boltzmann has uncovered the logical nucleus of the time concept.

But if time direction is a matter of probability and statistics, it requires reference to the ensemble of branch systems. The history of one isolated system as a whole does not supply a direction of time. And even a section of this history, an upgrade of the entropy curve, cannot answer the reversibility objection unless this section exhibits a structure designed by systems that branch off from the main system. The idea of countering the reversibility objection by reference to systems whose initial low-entropy states are produced by outside causes has been proposed repeatedly.[7] But such attempts have been usually regarded as unsuccessful, because of the widespread opinion that it

---

[7]For instance, this view was set forth by J. D. van der Waals, Jr., in "Über die Erklärung der Naturgesetze auf statistisch-mechanischer Grundlage", *Phys. Zeitschr.*, Bd. 12 (1911), pp. 547–549; and by Paul Hertz in his article "Statistische Mechanik", in *Ergebnisse der exakten Naturwissenschaften*, Bd. I (Berlin, published by the editor of *Naturwissenschaften*, 1922), p. 76. Similar considerations were presented in Hans Reichenbach, "Ziele und Wege der physikalischen Erkenntnis", in *Handbuch der Physik*, ed. by Hans Geiger and Karl Scheel, Bd. IV (Berlin, Julius Springer, 1929), p. 63. However, a symbolization by means of a probability lattice was not yet used in these publications.

must be possible to define a time direction by the use of one-system probabilities, that is, for isolated systems. That this aim cannot be fulfilled and that it represents an unnecessary restriction should now be sufficiently clear.

A statistical definition of time direction presupposes a plurality of systems which in their initial phases are not isolated, but acquire their initial improbable states through interaction with other systems, and from then on remain isolated for some time. That our universe, which is an isolated system, possesses a time direction, is due not merely to the rise of its general entropy curve, but to the fact that it includes a plurality of branch systems of the kind described. The direction of time is supplied by the direction of entropy, because the latter direction is made manifest in the statistical behavior of a large number of separate systems, generated individually in the general drive to more and more probable states.

• 16. The Hypothesis of the Branch Structure

Let us now investigate what assumptions are made when the existence of branch systems of the kind described is asserted. Such assumptions can be formulated in three ways: first, in a language L_1 employing usual time, the positive direction of which is defined by the direction of increasing entropy of the main system; second, in a language L_2, in which the direction of positive time, as defined, is opposite to that in L_1; and thirdly, in a neutral language L, which does not refer to a time direction.

Using language L_1, we regard section 1–6 of the curve of figure 21, in §15, as going from left to right, that is, as an upgrade; and we thus distinguish between *branch-off* points, such as 1, 3, and 5, and *turn-in* points, such as 2, 4, and 6. The branch-off points are points of relatively low entropy, that is, points at which the entropy of the subsystem is low. We shall call them *low-points*. The turn-in points are points of relatively high entropy; here the entropy of the subsystem is high, but that of the main system is still low, though a little higher than that at the corresponding branch-off point. We shall speak of these turn-in points as *high-points*.

In L_2 this terminology is changed, as follows. The branch-off points of L_1 become turn-in points, and the turn-in points of L_1 become branch-off points. With reference to section 6–1 of figure 21, we then call 6, 4, and 2 branch-off points; and 5, 3, and 1 are called turn-in points. This means that in L_2 we regard section 6–1 as going from right to left, and thus as a downgrade. With respect to section 7–12, or 12–7, we find opposite results. This section goes in L_1 from right to left, and in L_2 from left to right. A language L_i, therefore, is merely sectionally defined, and we do not extend it beyond its section.

What we call low-points and high-points, however, are the same points in both languages. For instance, 1, 3, and 5 are low-points in both L_1 and L_2; and 2, 4, and 6 are high-points in both languages. These terms, therefore, belong to the *neutral* language L. In L we do not speak of time, but only of entropy values; and we have here phrases like "low entropy" and "direction of growing entropy". The term "branch system" can also be used in L, when the distinction between branch-off points and turn-in points is not referred to. We then speak merely of the two ends, or terminals, of the branch system, and distinguish them as *lower end* and *upper end*.

When we wish to formulate the presuppositions made in assuming a branch structure of the entropy curve, it is dangerous to use a language in which a time direction is defined. When Boltzmann writes that a system of bodies entering into interaction is found, in general, in an improbable state, his statement appears plausible to us; but in the phrase "entering into interaction" the time direction of L_1 is assumed, and we must inquire whether the plausibility of the argument is not derived from tacit assumptions about time.

For this reason, we shall use the neutral language L to formulate the assumptions concerning the branch structure on which the statistical definition of time direction is based. These assumptions may be comprised under the name *hypothesis of the branch structure* and have the following form:

Assumption 1. The entropy of the universe is at present low and is situated on a slope of the entropy curve.

Assumption 2. There are many branch systems, which are isolated from the main system for a certain period, but which are connected with the main system at their two ends.

Assumption 3. The lattice of branch systems is a lattice of mixture as formulated in (12, 11–13), (12, 16), and (12, 18).

Assumption 4. In the vast majority of branch systems, one end is a low-point, the other a high-point.

Assumption 5. In the vast majority of branch systems, the directions toward higher entropy are parallel to one another and to that of the main system.

These five assumptions may be regarded as empirical hypotheses which are convincingly verified. We discussed the verification of assumption 1 in §15. The only possible objection to assumption 1 is that we are not certain whether such a thing as the entropy of the universe can be meaningfully defined. This objection was analyzed in §15. If such a cosmic entropy appears unacceptable, assumption 1 would have to be rephrased as indicated in §15; for instance, in the form: Every system is part of a larger system that is approximately isolated and has a rather low entropy. But assumptions 2–5 could still be maintained, assumption 5 with the modification that the phrase "and to that of the main system" would be omitted.

Boltzmann, it seems, believed that assumptions 2–5 can be derived from assumption 1. But this is obviously not true. In particular, assumption 5 is not derivable from assumption 1. If the entropy gradient of a branch system, or even of all of them, is counterdirected to that of the main system, such an arrangement does not contradict the laws of mechanics and is therefore not excluded by the causal definition of time order. This definition tells us that, for instance, in figure 21 the time direction from 1 to 2 is the same for the main system as for the branch system: either 1 is earlier than 2 for both systems, or later. But it does not follow that the entropy must change in the same way along both lines. In fact, for very small systems, such as those given by fluctuation phenomena, counterdirected changes do occur. In other words, the statement "the definition of time direction through entropy is consistent with the causal definition of time order" is an empirical statement. Assumption 5 may be called the *principle of the parallelism of entropy increase*.

Boltzmann's argument may be interpreted in the sense that a low entropy of the universe makes the existence of small systems of relatively low entropy probable. This assumes the existence of a sort of shuffling process; it asserts that inhomogeneities will be equally distributed in all smaller parts of the universe. But this consideration leads only to assumption 4, which, even if true, does not imply that the lower end of the branch system is situated in such a way that the entropy growth of the subsystem parallels that of the main system. The low-point may be situated at the other end of the branch. When we exclude this case by adding assumption 5, we introduce a distinct hypothesis. It is this hypothesis of the parallelism of entropy increase, hidden in Boltzmann's phrase "entering into interaction", which makes the low-point the beginning of reactions between the parts of the isolated system. But only because of assumption 5 are we entitled to interpret the low-point of the branch as the beginning of interaction, if the word "beginning" is meant to refer to the same time direction for all systems, including the main system and all subsystems. When we do not wish to refer to time direction, we call the low-point merely *interaction point*, leaving open the question whether the system here enters into its isolated development or finishes it. The high-point does not display interaction, because it is a point of equilibrium. Furthermore, the term may refer to interaction between the branch system and its environment, which interaction, too, assumes a greater intensity only at the low-point.

It is perhaps possible to show that in the history of a system, such as the universe, slopes satisfying assumptions 2-5 are much more frequent than slopes that do not satisfy these assumptions. Hence, we would have a high one-system probability that if assumption 1 is true, then assumptions 2-5 are true. But this probability would not help us very much, since it refers to a frequency realized in the very long time periods of the history of the universe. The assumption that we actually are on a slope of this kind would still be subject to empirical proof.

However, another proof can be given. If assumption 4 is accepted, it is possible to derive assumption 5 from assumption 3, that is, from the conditions holding for a lattice of mixture. It is therefore possible to omit assumption 5; the hypothesis of the branch structure is completely stated in assumptions 1-4.

Let us first explain this result with the help of an informal consideration. We write the lattice of branch systems in the time direction of L_1, that is, beginning with the terminals which in L_1 are branch-off points. The term "branch-off point" is well defined, because the time *order* of every branch system is defined through causal relationships, independently of the direction of its entropy increase. Let us now assume that about one-half of the branch systems had an entropy increase in the opposite direction, that is, that their branch-off points, in L_1, would be high-points, whereas their turn-in points, also in L_1, would be low-points. In the lattice (14, 1) the first column would then have equal numbers of states of low and of high entropy. Relation (14, 5), which formulates lattice invariance, would then still be true if A is a state of low entropy and B a state of high entropy; but this relation would be false if A is a state of high entropy and B a state of low entropy, at least for later columns. The reason is that for this latter case the probability on the right in (14, 5) is very low, whereas in the lattice which has equal numbers of states of low and of high entropy in the first column the corresponding vertical probability—that is, the expression on the left in (14, 5)—would be high, because practically all systems beginning (in L_1) with a high-entropy state would here develop toward low-entropy states.

These considerations can be transformed into a strict derivation of assumption 5 from assumptions 3 and 4. It was shown in §11 that, whatever be the distribution in the first column, later columns of a lattice of mixture assume distributions which converge to the distribution existing in the rows. Since a low-entropy state A possesses a low probability in a row, it must occur in later columns with a corresponding low frequency. It follows that only the initial columns of the lattice can possess a large number of low-entropy states; the later columns cannot. If the branch systems can be divided into two groups such that the systems of the first group begin with states of low entropy, whereas those of the second group end with low-entropy states, then it is not possible to combine these two groups into a lattice of mixture unless one group consists of a practically vanishing number of systems. Since assumption 4 tells us that there are a large number of systems having as one terminal a low-entropy state, we conclude that practically all these terminals must be on the same side; otherwise, the resulting lattice could not be a lattice of mixture satisfying the condition of lattice invariance. We have thus derived assumption 5 from assumptions 3 and 4.

In making this inference, we assume that the rows of finite length run through by the branch systems are long enough to realize the limit conditions which the theory lays down for infinite sequences; and likewise, that the columns are long enough—that is, that we have a sufficiently large number of branch systems. But assumptions of this kind are unavoidable whenever a statistical theory is applied to physical occurrences. It is a consequence of these assumptions of numerical adequacy that the derived conclusion does not represent a strict law for a given finite number of individuals; it is subject to possible exceptions. In order to illustrate this peculiarity let us imagine what we would observe in an exceptional case. Let us assume that among the many galaxies there is one within which time goes in a direction opposite to that of our galaxy. We would then have the situation envisaged by Boltzmann.[1] In this situation, some distant part of the universe is on a section of its entropy curve which for us is a downgrade; if, however, there were living beings in that part of the universe, then their environment would for them have all the properties of being on an upgrade of the entropy curve.

That such a system is developing in the opposite time direction might be discovered by us from some radiation traveling from the system to us and perhaps exhibiting a shift of spectral lines upon arrival. Because of the great distance, the message would reach us so late that it would merely inform us about the time direction which the system had aeons ago. We thus cannot acquire knowledge of a system that has, at the present time, a time direction opposite to ours, a limitation unknown to Boltzmann. This holds, at least, when we employ the usual

[1] See §15, n. 2.

definition of simultaneity according to which light travels in both directions along its path with equal speed. When we abandon this definition, the aeons of time could be transformed into a very short time period. The indeterminacy of simultaneity leaves the time comparison between the two systems indefinite within wide limits (see 5, 3). Furthermore, the radiation traveling from the system to us would, for the system, travel backward in time, that is, would not leave that system but arrive at it. Perhaps the signal could be interpreted by inhabitants of that system as a message from our system telling them that our system develops in the reverse time direction. We have here a connecting light ray which, for each system, is an arriving light ray annihilated in some absorption process; judged from the other system, it is therefore emitted at its source by an irreversible process going in reverse.

Such possible exceptions to entropy parallelism do not affect the given derivation of assumption 5 from assumptions 3 and 4, because assumptions 3 and 4 refer to a large number of systems. Statistical hypotheses are not disproved by individual exceptions. It may be added that up to the present time no kind of astronomical evidence has been found which indicates the existence of such exceptions.

It should be noted that the given derivation of assumption 5 does not compel us to use language L_1. Using L_2, we can construct a lattice satisfying assumption 5 by putting into the first column all the high-points of the branch systems. The systems would then develop toward low-entropy states. For this lattice, the condition of lattice invariance would be satisfied in reverse, so to speak; we merely replace "$+m$" by "$-m$" in the expression on the left in (14, 5). But the physical fact that all branch systems change their entropies in the same direction would also hold for L_2, being independent of the choice of the language. Indeed, the lattice thus constructed is the same as the one constructed for L_1, if the latter is read from right to left. That we prefer to use L_1 and thereby identify positive time with the increase of entropy is a matter of definition, logically speaking. The psychological grounds for this definition will be pointed out later.

Since the condition of lattice invariance expresses the inference from time ensemble to space ensemble, it follows that, if this inference is applicable, practically all branch systems must display an entropy increase in the same direction. A world in which mixing processes are governed by the relations holding for a lattice of mixture, formulated in (12, 11-13), (12, 16), and (12, 18), exhibits a certain uniformity; we will call it *statistically isotropic*. The term refers to the fact that the distribution of the time ensemble is repeated in the space ensemble whenever a mixing process has been going on for some time, as formulated in relation (12,10). Time direction and space direction display here a certain equivalence which, by analogy with a terminology used in mechanics and optics, may be called an *isotropy*. We thus find that branch systems define a unique direction of time because the world is statistically isotropic.

Although we have thus derived assumption 5 from assumptions 3 and 4, we must not forget that both assumptions 3 and 4 formulate physical properties of the world and are by no means logically necessary. With respect to assumption 4, this is fairly obvious; but it applies to assumption 3 also. The condition of lattice invariance is not derivable from the axioms of the calculus of probability, but represents a separate addition, the applicability of which is subject to empirical test (see §12). All available evidence, of course, strongly supports assumption 3. We may say, therefore, that the world owes its time direction to a certain uniformity exhibited by all its mixing processes, to its statistical isotropy. The direction of time reflects a general statistical trend, the trend to portray the time ensemble in the space ensemble. This trend controls the ensemble of branch systems as well as any ensemble of molecules which constitutes an individual branch system. The parallelism of entropy increase expresses a general statistical law holding for this universe, the same law that controls the mixing of molecules within a gas system.

The physical law expressed in the statistical isotropy of the universe, and consequently in the parallelism of entropy increase, does not have the logical character of a causal law. It may be called a *cross-section law* of the universe, that is, a law exhibited in the structure of a simultaneity section, a section for which t is constant, extending across the universe, or, rather, in the comparison of the structures of several such cross sections. Laws of this kind play an important part in physics, though their number seems to be small.[2]

One might attempt to explain the parallelism of the branch systems as causally forced upon them by their being not completely insulated from the universe; the small remaining coupling to the main body of the universe would then be regarded as supplying the transmission of a causal influence which makes all the branch systems adjust themselves to the main system. But such an interpretation would not be

---

[2] Another law of this kind is Heisenberg's relation of indeterminacy, which follows from the cross-section law that all observational material can be summed up in a ψ-function; see Hans Reichenbach, *Philosophic Foundations of Quantum Mechanics* (Berkeley and Los Angeles, Calif., Univ. Calif. Press, 1944), p. 4. This book will be cited hereafter as *PhF*.

tenable, because the statistical isotropy would also result if the systems were completely insulated from one another.

To show this, suppose that each horizontal line of the lattice (14, 1) represents the history of a completely isolated branch system, extending from negative infinite time to positive infinite time. The lattice would then be infinite both to the right and to the left. Assume further that each line can be shifted horizontally and that we arrange the lines in such a way that one vertical column contains only low-entropy states. Then this column would divide the lattice in two halves, each of which would exhibit, according to our physical knowledge, the relations of a lattice of mixture and thus of statistical isotropy. In order to verify this result, we would count the right-hand half going to the right, and the left-hand half going to the left. The lattice would thus show a directional symmetry, like the entropy curve of a closed system. But if we cut off one half of this lattice, the remaining lattice would display a direction in its initial columns. The fact that branch systems, at their low-points, are connected with the main system and do not exist, according to language L_1, before this point, constitutes the physical analogue to the mathematical operation of cutting off the lattice. Note that corresponding halves of the horizontal lines are defined in terms of order relations and do not presuppose direction; that is, we cannot reverse one of the lines within the lattice while leaving the others unchanged, because this reversal would violate order relations.

Statistical isotropy is thus a law governing causally independent systems; it expresses the fact that such systems, in their spatial distribution, satisfy conditions of randomness. If we threw ten thousand dice simultaneously, we should observe that their six faces turn up in about equal numbers of cases. This satisfaction of the condition of isotropy is not a consequence of the weak coupling of the events in terms of molecular collisions with their environment. Although we can never observe completely independent events, such a conclusion can be drawn as a limit statement derived for experimental arrangements in which the causal isolation is made better and better. And the random distribution is not merely a definition of causal isolation, because the isolation can be tested by direct observational methods.

When we call statistical isotropy a cross-section law, we wish to express by this term the absence of causal connection. A cross-section law must therefore be distinguished from causal correlations within a cross section, which are produced by common causes and which will be studied in §19. For instance, when we consider a cross section of the population of the United States, we find that the great majority of persons speak English; this is a causal correlation, not a cross-section law. That our universe is governed by certain cross-section laws, in addition to causal laws, is a matter of fact; and we must not misinterpret this fact and assume as an a priori necessity that independent systems follow the laws of randomness. All we can do is formulate the existing law and test it by means of observations.

Let us summarize these considerations. The statistical definition of time direction is based on the hypothesis of the branch structure, formulated in assumptions 1–5. It is merely *one* of these assumptions that the universe is at present on an upgrade of its entropy curve; and it is necessary to qualify this by some of the other assumptions referring to the existence of branch systems. These include an assumption expressing the parallelism of entropy increase (assumption 5); but this assumption is derivable from the other four assumptions and thus a logical consequence of a general property of the universe called statistical isotropy and formulated in assumption 3. The set of assumptions 1–5 is therefore equivalent to the set of assumptions 1–4. This set of assumptions, the hypothesis of the branch structure, is not derivable either from the axioms of the calculus of probability or from the ergodic hypothesis (explained in §10). The latter hypothesis merely supplies the probability metric of the universe. In order to derive the directional properties of time, we have to add the *factual* assumptions 1–5. Our universe is at present on a section of its entropy curve satisfying assumptions 1–5: this is the fact to which we owe the existence of a direction of time.

The logical function of the assumptions 1–5 in the probability theory of time may be described as follows. In the ensemble of systems of our environment, we find a large number of isolated systems in entropy states and transitions which in the history of one system have a low probability. How can we explain the high many-system frequency of such isolated states of order in view of their low one-system probability? How can we explain the fact that such states are often preceded by states of even lower entropy? These questions are answered by reference to assumptions 1–5. We explain the frequent occurrence of isolated states of order by deriving it from the frequent occurrence of interaction points, which in turn is stated in assumptions 1–5. Explanation means logical derivation from the fundamental properties of the physical world; therefore, interaction explains order, and not vice versa.

These logical relationships are extremely important for the understanding of a problem which must now be investigated: the distinction between cause and effect. This investigation leads us to another chapter of our inquiry.

CHAPTER IV THE TIME DIRECTION OF
MACROSTATISTICS

● 17. Macroarrangements and Macroentropy

We shall now extend our statistical considerations to processes other than mixtures of molecules, namely, to processes the elementary "particles" of which are macroscopic objects (such as grains of sand or playing cards) and the elementary arrangements of which are not microstates, but macrostates. For processes of this kind, classes of arrangements can also be defined; such a *class of arrangements* may be distinguished from the *arrangements* themselves by being designated as a *state*. The previous results concerning probability can be applied to these states when we count the number of arrangements within the class; and the probability of the state then assumes a function corresponding to entropy, being low for ordered states and high for states of disorder. We may speak here of the statistics of *macroarrangements*, and shall also use the terms *macroprobability* and *macroentropy*. We thus arrive at statistical relationships which refer exclusively to the macrocosm and which may be called *macrostatistics*, in contrast to the *microstatistics* of the kinetic theory of matter. These macrostatistics reveal the existence of a direction analogous to that of thermodynamic processes.

We begin by defining the concept of order for macrostatistics. It was shown above (§8) that for microstatistics the concept of *order* is reducible to that of *same order*, which concept, in turn, is definable in terms of the concept macroscopically indistinguishable. This last-mentioned concept cannot be applied to macroarrangements, because all such arrangements are distinguishable. When we shuffle a deck of cards, for example, each arrangement of the cards existing at a certain moment is distinguishable from every other arrangement; we can identify it, if we like, by writing down the names of the cards in the order they happen to have. In what sense, then, can we speak of the same order of macroarrangements?

With reference to the example of the deck of cards, we may call attention to the fact that we do distinguish between ordered and unordered arrangements. If all red cards are on top and all black cards below them, we speak of an ordered arrangement, whereas we regard the kind of arrangement usually resulting from shuffling as unordered. The reason is that the first kind of arrangement can be characterized by a simple rule, whereas the latter kind cannot.

It would be a mistake to believe that an unordered arrangement satisfies no rule whatsoever. Since the number of cards is finite, we can always find arithmetical rules fitting a given arrangement. For instance, let us write the number 1 for a red card and the number 2 for a black card, putting a decimal point before this series of numbers. We then obtain a decimal fraction of 52 decimal digits, which is easily converted into a fraction of two integers m/n. The given arrangement of cards is therefore described by the rule that it corresponds to the decimal expansion of the fraction m/n. But this is not a simple rule; the computation of the fraction m/n requires involved logical operations, which certainly cannot be performed without pen and paper.

A rule may be called simple if we can see at first sight whether it is satisfied by a given arrangement. The rule "All red cards are on top" and the rule "Every second card is red" are of this kind. This is certainly not a very precise definition of the term "simple"; and if a rule is not too simple and yet not too complicated, we may disagree whether it still is to be included among simple rules. However, for many practical purposes the definition is precise enough. It is one of those definitions that portray language behavior sufficiently without pretending to introduce exactness into a domain where it is inappropriate. It makes the phrase "simple rule" a psychological term. When we speak of arrangements that can be described by simple rules and others that cannot be so described, therefore, we use a classification in terms of reactions by human observers.

If we wish, we may even classify observers into different categories. An expert at bridge will use a rule such as "a set of cards promising success", or "a good hand", differently from a poor bridge player. A grain merchant will classify samples of grain into quality groups which the untrained eye cannot distinguish. We may therefore speak of simple rules relative to a certain category of observers. For general statistical considerations, however, it does not make much difference to what kind of observer we refer.

We will call two arrangements a_1 and a_2 *rule-indistinguishable* if every simple rule satisfied by a_1 is also satisfied by a_2, and vice versa. When we shuffle a deck of cards, most resulting arrangements are rule-indistinguishable in the following sense: they satisfy the simple rule, "being an arrangement of one deck of cards", but no other simple rule. In contrast, an arrangement satisfying the specification "all red cards on top" is rule-distinguishable from random arrangements, because it satisfies this simple rule in addition to the general rule holding for all arrangements. The arrangements satisfying this particular rule still constitute a large number, since we can shuffle the red cards and the black cards separately. Among these, there are arrangements satisfying more specific rules, such as the rule "first all diamonds, then all hearts, then all spades, then all clubs". We can speak, however, of those arrangements that satisfy merely the rule "red cards on top" and do not satisfy any further simple rule; these arrangements are rule-indistinguishable from one another.

Two arrangements that are rule-indistinguishable are said to have the *same order*. For a given arrangement we can thus define the same-order class, which we call the *state* of the arrangement, and can take over into our macrostatistics the methods of microstatistics. The degree of order of an arrangement is given by the extension of its same-order class (cf. §9).

This definition presupposes the assumption that all arrangements are equiprobable. For shuffling cards this is certainly correct. For other applications this assumption of a probability metric must be tested empirically, since we do not possess analogues of Liouville's theorem for them. It may sometimes turn out that we have to introduce a different metric by attaching different statistical weights to the arrangements. A crude estimation of the metric is often sufficient.

According to these definitions, the shuffling of a deck of cards represents a macromechanism which shows all the statistical properties of micromechanisms. An ordered arrangement of cards has a low macroprobability, the logarithm of which may be called the macroentropy of that arrangement. The act of shuffling transforms an initially ordered arrangement into unordered arrangements, which have high macroprobabilities, and hence high macroentropies. The shuffling thus far corresponds to the first section of the curve of figure 19, §13. During further shuffling the curve will remain on a high level, going through slight fluctuations. But the occurrence of a long downgrade is very improbable. The shuffling process thus has the properties of an irreversible process; it may be called *quasi-irreversible*. It differs from irreversible processes in the proper sense because we can assort the shuffled cards and thus bring them back to their original order by individual selection, an act which is impossible for microstatistics.[1] But it is highly probable that we cannot return them to their order by shuffling.

The shuffling of cards is a rather involved mechanical process, in which the result of each individual step can be predicted with probability only. It may therefore be of some interest to construct a quasi-irreversible process for which the shuffling mechanism consists in simple mechanical processes of an openly deterministic character. Suppose we arrange a set of billiard balls in a straight row along the middle line of a billiard table (fig. 22), leaving short distances between the balls. We then push another ball, lying outside the row, in such a way as to hit one ball of the row in a slanted direction. After the collision, the two balls will hit the walls and, being reflected, hit other balls. When we let this process go on for a few minutes, the ordered arrangement of the balls will be destroyed, and all the balls will be arranged irregularly on the table. This process, in which order is transformed into disorder, is quasi-irreversible in the sense that it is impossible for us to reverse it. Of course, if we could stop all the balls simultaneously in the unordered state and push them with equal

---

[1]This problem has been discussed repeatedly with reference to Maxwell's daemon, who assorts molecules with respect to speed by opening a sort of sash window when a fast molecule arrives and closing it on the arrival of a slow one. It has been shown that the opening and closing of the window is connected with a rise in entropy, and that thus even a daemon whose size does not exceed atomic dimensions cannot violate the second law of thermodynamics. See Leo Szilard, "Über die Entropieminderung in einem thermodynamischen System bei Eingriffen intelligenter Wesen", *Zeitschr. f. Physik*, Bd. 53 (1929), pp. 840–856.

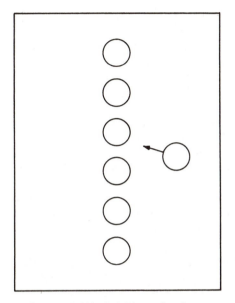

Fig. 22. Billiard balls ordered in a row. If a ball is pushed from the side so as to collide with one of the balls in the row, this and subsequent collisions destroy the order of the balls. The order will not be restored by chance while the balls are rolling across the table.

speeds in exactly opposite directions, the process would go in reverse, and the original ordered arrangement would result. But it is technically impossible for us to do so.

This imaginary experiment is the exact analogue of the reversal of molecular speeds discussed in §13. The unordered arrangement in which we reverse the speeds possesses an *implicit order*, which escapes our attention; this arrangement would be rule-indistinguishable from other unordered arrangements. This illustration shows that macrostatistics is controlled by relations similar to those holding for microstatistics and can be used to define a time direction in the same sense as is possible for Boltzmann's statistics of molecules. For instance, when we find the billiard balls neatly arranged in a row, or the deck of cards ordered in such a way that the red cards are on top, we conclude that this state was produced by intervention. It is too improbable that such an ordered state should result by chance while the balls are rolling, or the cards being shuffled.

• 18. Cause and Effect: Producing and Recording

There exists an essential difference between microprocesses and macroprocesses. The former possess a natural shuffling mechanism, given by the collisions of molecules. The latter often do not possess any natural shuffling mechanism and must therefore be artificially shuffled, as is done with a deck of cards; in other processes, the natural shuffling mechanism is so very slow that, at a given moment, the system remains practically unchanged, as, for instance, sand which is slowly shuffled by the wind. This distinctive feature leads to peculiar consequences for macrostatistics: states of high order can here be preserved for a long time and can be observed conveniently. This is the reason why macrostatistics supplies what we call *records* and why, at the same time, it presents us with the key to the understanding of *causal explanation*. This peculiarity must now be studied in more detail.

Suppose we find in the sand traces of footprints, somewhat smoothed out by the wind, but still recognizable as impressions of human feet. We conclude from this "record" that at some earlier time a man walked over the sand, thus causing the footprints. What is the logical schema of this inference?

The different arrangements of grains of sand can be classified into states by the use of simple rules. One state is given by a smooth surface of the sand; another one by the condition described, in which the surface carries the imprints of human feet. When we wish to compute the probabilities of such states it would be incorrect to assume that all arrangements of grains of sand are equiprobable. While the sand is left alone, exposed only to the shuffling by the wind, its arrangements are slowly changed; but the wind exerts a discriminating influence, blowing away little mounds of sand and filling holes, with the effect that a smooth surface, or a surface showing wavelike patterns, is more probable than a state in which the sand shows distinct holes. This is the probability metric which observation teaches us to hold for the isolated sand system, when we understand by isolation the exclusion of all influences other than the agitation by the wind or the waves of the ocean.

In terms of this metric, the state given by the footprints is a highly ordered state, whereas a smooth surface is an unordered state; in fact, the latter is the state of equilibrium for the sand. We have here the case that a highly ordered state is preserved for some time; and we ask, how can we explain the presence of this ordered state?

This is the same type of question which was discussed above (§14) with respect to microprocesses. We observe a state which in the history of the isolated system is very improbable; how can we account for it? The answer is, as before, that we assume the observed state to be a product of interaction, that we prefer an interpretation in which the system was not isolated in the past. And as for thermodynamical statistics, this inference is based on a many-system probability, in contract to a one-system probability. It therefore presupposes the hypothesis of the branch structure, applied, however, to statistics of macroarrangements.

For this reason, the footprints take over the function of a record. They allow us to infer that at some earlier time an interaction took place, that a person's steps caused the ordered state of the sand; because this state was not "shuffled away", it is a record of the interaction. Records are ordered macroarrangements the order of which is preserved; they are frozen order, so to speak.

Static order does not contradict assumption 3 of §16, since that assumption refers to molecular arrangements and the system remains merely macroscopically unchanged. The thermic state of the system goes through its inevitable development. While the foot pushes the sand grains, they pick up kinetic energy, which is transformed into heat when the grains come to rest. The history of the footprints represents a branch going from a low-point to a high-point and ending in a state of thermodynamical equilibrium. But the macroscopic equilibrium has not yet been reached; and in the sense of macrostatistics and macroentropy, the branch system has not completely run its course and still displays order. However, the stirring action of the wind slowly steers the system toward a state which even for macrostatistics represents an equilibrium, or a state in which the traces of macro-order are erased and the footprints are completely smoothed out.

In addition to the clarification of the nature of records, the example of the footprints also helps us to analyze the meaning of *causal explanation*. Explanation in terms of causes is required when we meet with an isolated system displaying a state of order which in the history of the system is very improbable. We then assume that the system was not isolated at earlier times: explanation presents order in the present as the consequence of interaction in the past. The otherwise improbable occurrence is thus presented as probable, because the low one-system probability is replaced by a high many-system probability. When we wish to explain improbable order in terms of causes, we follow the rule that explanation consists in deriving the improbable from the branch structure of the universe (see assumptions 1–5, §16).

These results were developed by the use of language L_1. Translating them into neutral language L, we arrive at the following explication of the terms *cause* and *effect*: *The cause is the interaction at the lower end of the branch run through by an isolated system which displays order; and the state of order is the effect.*[1]

There are many illustrations of this interpretation of causal explanation, which, when we go back to language L_1, makes the word "cause"

---

[1][This explication of "cause" and "effect" cannot be considered completely general; ordinary usage gives these terms a much broader application. For an explanation of the manner in which an extended meaning arises out of this narrow meaning, see p. 156.—M. R.]

a term referring to the past. A scar in the human skin is regarded as caused by the incision of a knife; puddles of water found high in the mountains in a dry summer are regarded as caused by melting snow; when we find two pieces of paper having irregular edges which fit together, we conclude that they were produced by an act of tearing one piece of paper in two; ink marks on paper are explained by an interaction, the act of writing, which caused the ordered lines on the paper, and so on. In all these examples, the order observed is an order of macroarrangements which was preserved after the system, subsequent to the interaction, had reached its thermodynamic equilibrium. Here is the significance of macrostatistics: it supplies us with criteria of order which escape microstatistics, and it helps us to discover causal chains in the macrocosm. In fact, the origin of the concept of cause is presumably to be sought in experiences in which man has been confronted by unexpected states of macro-order which have stimulated his imagination to assume past causes. But the concept applies likewise, of course, to states of ordered microarrangements, since these states also call for explanation in terms of interaction, as was explained in §14.

Sometimes the order of the macroarrangement is not easily ascertainable and specific training may be required for its discovery. Records of this kind, which only the expert observer knows how to read, play an important part in geology and anthropology. We may mention fossils in the ground, caused by living organisms' being buried in the sand; stone tools cut by prehistoric men; granite rocks found arranged in relatively ordered lines, deposited at the melting edges of glaciers in earlier geological periods. With respect to the last example, it took years to discover that the arrangement of the rocks was ordered and called for explanation. Sometimes a slow change in the macroarrangement can be used to infer the length of time that has passed since the interaction. Of this kind is the ascertainment of the age of rocks and other bodies from the ratios of products of radioactive decay contained in them. In other cases, the change consists merely in an accretion of further material, as in the rings of trees, which are caused by the yearly deposition of layers of cells and thus record the age of the tree.

It may be recalled that our analysis has thus far been given by the use of language L_1. Only the definition of cause and effect was given in a different language, namely, in the neutral language L. We found that when we use this definition, explanation in L_1 must always refer to the past. Let us now check our results by using language L_2, which employs the opposite time direction.

For this purpose, we will study the example of the footprints in the sand by the use of language L_2. Imagining the history of the footprints in reverse time, we come to the following description. First, there are smoothed-out imprints in the sand, somewhat resembling the shape of a human foot. Then winds begin to blow, carrying grains of sand back and forth, with the effect that the shapes in the sand become more distinct and eventually are formed into the exact molds of human feet. Finally a man arrives, walking backward. He puts his feet one after the other into the imprints in the sand, which fit his feet perfectly. When he lifts his foot, sand pours in from all sides and fills the imprints completely, so that this spot of sand no longer differs from its environment.

In this interpretation, the order of the smoothed-out footprints is not regarded as a record; it does not follow, but precedes, interaction, and it represents not a *postinteraction* state, but a *preinteraction* state. The very improbable happenings to which it leads can here not be explained in terms of past occurrences. When we pursue the impacts of the wind into the past of L_2, we might argue that the particular condition of the particles of air was of such a kind as to lead to the unusual series of impacts which transform the smoothed-out imprints into shapes exactly fitting a man's feet. But then we have explained one state of order in terms of another one, given by the condition of the particles of air. Explanation requires reducing order to interaction; and no such explanation can be found in the past of L_2.

We would therefore come to a very different conclusion. We would explain the improbable coincidences by their *purpose* rather than by their *cause*. The wind transforms the molds in the sand into the shapes of human feet *in order that* they will fit the man's feet when he comes. The man steps exactly into the imprints *in order that* they will vanish when he lifts his foot. When we use language L_2, in which time direction is opposite to the usual one, explanation leads to *finality* instead of *causality*. The reason is that, as in L_1, explanation consists in deriving the improbable order from the branch structure of the universe formulated, in neutral language, in assumptions 1–5, §16. Explanation in L_2, therefore, requires reference to future interaction, which is the intermediate stage in the derivation from assumptions 1–5. The definition of cause, given earlier in the present section, could be taken over into the vocabulary of L_2 simply by the substitution of the word "purpose" for the word "cause": the future interaction, which explains the improbable order of the present, would be called the purpose of the occurrence.

This kind of explanation appears implausible to us; yet it is merely the translation of our own experiences into a language of the opposite time direction. We see that our conception of *causality*, of the past that determines the present and the future, is closely connected with our definition of positive time in terms of growing entropy. In opposite time we find its equivalent in a conception of *finality*, according to which the future determines the present and the past.

Which is the correct language, that of causality or that of finality? This is certainly a meaningless question. The two languages L_1 and L_2 represent *equivalent descriptions* (see §4); one is as true as the other. That L_1 appears to us as a more natural language, that we are so strongly disposed toward the identification of the direction from interaction to order with positive time, has its basis in the nature of the human organism. The discussion of this problem may be postponed.[2]

We conclude: If we define the direction of time in the usual sense, there is no finality, and only causality is accepted as constituting explanation. This point may be made clearer by the following considerations. If someone argues that it is a matter of convention to select the direction of growing entropy as the direction of time, his conception cannot be called false. But he must not commit the error often connected with other forms of conventionalism: the error of overlooking the empirical content associated with the use of this convention. It is an empirical fact that in all branch systems the entropy increases in the same direction. For this empirical reason, the convention of defining positive time through growing entropy is inseparable from accepting causality as the general method of explanation. Those who prefer to give an explanation in terms of finality would be compelled to use the opposite time direction and to regard time as going from high-entropy states to low-entropy states. Then, however, finality is but another language describing the same universe, whereas the physics formulated in this language would not differ from the present one. The use of such a language, of course, would be extremely inconvenient, because it contradicts the time direction of psychological experience.

This analysis shows that the use of growing entropy for the definition of time direction leads to the consequence that causes, and not ends, determine the occurrences of the present. When the scientist refuses to accept teleology as a mode of explanation, he has good reasons for doing so: it is the second law of thermodynamics which excludes finality for a world like ours, in which growing entropy defines positive time. Once the time direction is assumed in the usual sense, it is not a matter of personal preference, not a mode of consideration, whether we should describe the world in terms of causes or of ends: it

---

[2][This problem would have been discussed in the projected chapter on the human mind, which was to have been the final chapter.—M. R.]

is a physical law that causality, and not finality, governs the universe. The link between growing entropy and cause as the determining factor of all happenings is given by the statistical isotropy of the universe, which makes all branch systems develop in the same direction and places the point of interaction at that end of the branch which corresponds to a lower-entropy state of the universe. For the time direction of growing entropy, therefore, the interaction point is the beginning, not the end, of the evolution of the branch system. The statistical relationships developed in the lattice of mixture account for our conception that the past *produces* the future, and not vice versa.

The word "produces" is a statistical concept. It refers to the direction from interaction to isolated states which still display order; and it means that the large number of isolated systems found in states of order is a logical consequence of the frequent occurrence of interaction states in which systems branch off at a relatively low entropy. The interaction state is thus regarded as the cause of the ordered states. The distinction between cause and effect is revealed to be a matter of entropy and to coincide with the distinction between past and future. Yet it presupposes not merely the increase of the entropy of the universe, but the existence of a branch structure satisfying assumptions 2–5, §16.

The cause produces the effect; but since the cause leaves traces in the effect, it can be inferred from its effect. Here is the origin of the distinction between past and future referred to in statement 5, §2. The statement that although the past can be recorded, the future cannot, is translatable into the statistical statement: *Isolated states of order are always postinteraction states, never preinteraction states*. This statement follows from assumption 5 if positive time is defined as the direction of increasing entropy. Examples of traces—of records of the past—have been given in previous illustrations.

The present is the intermediary between past and future; it contains the active agent that produces the future, and it contains the records of the past, which is completed and irretrievable. When we strip these terms of their emotional connotations and translate them into statistical language, we find that they refer to the transition from states of interaction to isolated states of order. We saw that the states of interaction have a logical priority over the isolated states of order because their frequent occurrence is formulated in assumptions 1–5 and is thus regarded as a fundamental property of the universe, from which the frequent occurrence of isolated states of order is logically derivable. This logical priority can be regarded as the source from which the conception of producing as an activity springs; what is logically prior appears as the determining factor. With the meaning of "producing", that of the word "recording" is given, since it follows from the given statistical considerations that the relation of *recording* is the converse of the relation of *producing*. The cause produces the effect, the effect records the cause. If producing is regarded as an activity, recording therefore indicates passivity. These emotive terms thus find an explication in the statistical definition of time direction.

The given definition of cause refers to a specific relationship holding within a branch system. From this original meaning, the concept of cause has been extended to wider usage. We saw in §5 that an order of time can be defined (without reference to a direction) by means of causal chains, and that, furthermore, the causal net is ordered as a whole. It follows that if we assign the direction of positive time to one causal chain, such a direction is assigned to all causal chains. By this relationship, the direction of time is determined for any two events that are connected by a causal chain, even if the process is reversible.

The widest definition of cause and effect is therefore given by reference to time direction, which in turn is definable in terms of the narrower meaning of "producing". We can thus even speak of cause and effect with respect to reversible processes, calling the earlier state the cause of the later one. For instance, we say that the present position and velocity of the earth in its orbit around the sun is the cause of its position and velocity after three days; and we would not be willing to reverse this relationship. If the word "cause" used here merely referred to the motion of the earth around the sun, it would be meaningless to distinguish between cause and effect, since either one could be said to "produce" the other. But in the sense of a transfer from relationships holding for irreversible processes in branch systems, the use of the word "cause" is legitimate in application to macroprocesses governed entirely by the laws of mechanics.

One might also adduce the argument that, strictly speaking, no macroprocess is reversible, because every such process is connected with some thermic change and thus with entropy increase. For instance, the earth is slowed down in the rotation around its axis. In this sense, every macroprocess per se has a time direction. But this kind of entropy increase is not noticeable; and, psychologically speaking, the cause-effect distinction, in its application to mechanical processes, must be regarded as resulting from transfer.

• *19. The Principle of the Common Cause*

We shall now study some further applications of macrostatistics which lead once more to the distinction of cause and effect and thus to the definition of a time direction. The results will be formulated in a specific principle governing macroscopic arrangements. It will be shown, however, that this principle does not represent a new assumption, but is derivable from the second law of thermodynamics, if this law is supplemented by the hypothesis of the branch structure.

Suppose that lightning starts a brush fire, and that a strong wind blows and spreads the fire, which is thus turned into a major disaster. The coincidence of fire and wind has here a common effect, the burning over of a wide area. But when we ask why this coincidence occurred, we do not refer to the common effect, but look for a common cause. The thunderstorm that produced the lightning also produced the wind, and the improbable coincidence is thus explained.

The schema of this reasoning illustrates the rule that the improbable should be explained in terms of causes, not in terms of effects. Let us postpone an investigation concerning the relationship between this inference and the hypothesis of the branch structure, and let us first study the inference in its own right. The logical schema which governs it may be called the *principle of the common cause*. It can be stated in the form: *If an improbable coincidence has occurred, there must exist a common cause.*

In our daily life we often employ inferences of this kind. Suppose both lamps in a room go out suddenly. We regard it as improbable that by chance both bulbs burned out at the same time, and look for a burned-out fuse or some other interruption of the common power supply. The improbable coincidence is thus explained as the product of a common cause. The common effect, the fact that the room becomes completely dark, cannot account for the coincidence. Or suppose several actors in a stage play fall ill, showing symptoms of food poisoning. We assume that the poisoned food stems from the same source—for instance, that it was contained in a common meal—and thus look for an explanation of the coincidence in terms of a common cause. There is also a common effect of the simultaneous illness of the actors: the show must be called off, since replacements for so many actors are not available. But this common effect does not explain the coincidence.

Chance coincidences, of course, are not impossible: the bulbs may burn out simultaneously, the actors become ill simultaneously for different reasons. The existence of a common cause is therefore in such cases not absolutely certain, but only probable. This probability is greatly increased if coincidences occur repeatedly. Suppose two geysers which are not far apart spout irregularly, but throw up their columns of water always at the same time. The existence of a subterranean connection of the two geysers with a common reservoir of hot water is then practically certain. The fact that measuring instruments such as barometers always show the same indication if they are not too far apart, is a consequence of the existence of a common cause— here, the air pressure. Further illustrations would be easy to find.

It will be advisable to treat the principle of the common cause as a statistical problem. For this purpose we assume that A and B have been observed to occur frequently; thus it is possible to speak of probabilities $P(A)$, $P(B)$, and $P(A.B)$ with reference to a certain time scale. For instance, we may count the eruptions of the geysers within a scale of days, or hours, and thus arrive at the probability that the geysers will spout on a given day, or at a given hour. We shall at the same time generalize the problem by extending it to statistical relationships between cause and effect, that is, to situations in which the cause produces the effect only with a certain probability. The special case that this probability is practically equal to 1, as in the illustrations given, is then included in the general treatment.

The statistical relationships which two simultaneous events have, on the one hand, to a common cause, and, on the other hand, to a common effect, can be presented by the schema of figure 23. We have here a double-fork arrangement in which A and B represent the two events the simultaneous occurrence of which is improbable; C is their common cause; E, their common effect. We said that C, and not E, explains the simultaneous occurrence of A and B. Let us therefore omit the upper part of the diagram and study the single fork of figure 24, in which merely the common cause C is indicated.

That the simultaneous happening of A and B is more frequent than can be expected for chance coincidences, can be written in the form

$$P(A.B) > P(A) \cdot P(B) \ . \tag{1}$$

When we apply to the left side the general multiplication theorem of probabilities,

$$P(A.B) = P(A) \cdot P(A, B) = P(B) \cdot P(B, A) \ , \tag{2}$$

we derive from (1) the two relations

$$P(A, B) > P(B) \ , \tag{3}$$

$$P(B, A) > P(A) \ . \tag{4}$$

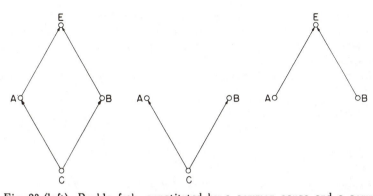

Fig. 23 (left). Double fork, constituted by a common cause and a common effect.

Fig. 24 (center). Fork open toward the future, constituted by a common cause.

Fig. 25 (right). Fork open toward the past, constituted by a common effect.

Each of these relations, vice versa, can be used to derive (1). Therefore (3), or (4), is equivalent to (1). In these derivations, it is always assumed that none of these probabilities vanishes.

In order to explain the coincidence of A and B, which has a probability exceeding that of a chance coincidence, we assume that there exists a common cause C. If there is more than one possible kind of common cause, C may represent the disjunction of these causes. We will now introduce the assumption that the fork ACB satisfies the following relations:

$$P(C, A.B) = P(C, A) \cdot P(C, B) \ , \tag{5}$$

$$P(\bar{C}, A.B) = P(\bar{C}, A) \cdot P(\bar{C}, B) \ , \tag{6}$$

$$P(C, A) > P(\bar{C}, A) \ , \tag{7}$$

$$P(C, B) > P(\bar{C}, B) \ . \tag{8}$$

It will be shown presently that relation (1) is derivable from these relations. For this reason, we shall say that relations (5)–(8) define a *conjunctive fork*, that is, a fork which makes the conjunction of the two events A and B more frequent than it would be for independent events. When we say that the common cause C *explains* the frequent coincidence, we refer not only to this derivability of relation (1), but also to the fact that relative to the cause C the events A and B are mutually independent: a *statistical dependence* is here derived from an independence. The common cause is the connecting link which transforms an independence into a dependence. The conjunctive fork is therefore the statistical model of the relationship formulated in the principle of the common cause.

For the mathematical treatment of the fork we introduce the following notation:

$$P(A) = a \; , \quad P(B) = b \; , \quad P(C) = c \; , \quad P(A.B) = w \; ,$$

$$(9)$$

$$P(C, A) = u \; , \; P(C, B) = v \; , \; P(\bar{C}, A) = r \; , \; P(\bar{C}, B) = s \; .$$

The rule of elimination[1] furnishes the following equations:

$$w = cuv + (1 - c)rs \; , \tag{10}$$

$$a = cu + (1-c)r \; , \quad b = cv + (1-c)s \; . \tag{11}$$

From these relations we find by the help of some simple computations the factorized form

$$w - ab = c(1-c)(u-r)(v-s) \; . \tag{12}$$

We will assume that $0 < c < 1$. Because of relations (7)–(8), the expression on the right is greater than 0; therefore $w > ab$. This proves relation (1).

We see, furthermore, that $w = ab$ if either $u = r$ or $v = s$, or both. This means that the statistical dependence of A and B derives from the inequality of the probabilities of A and B relative to C and to $\bar{C}$, formulated in (7)–(8). If $u < r$ and also $v < s$, we find, once more, $w > ab$; in this case, C and $\bar{C}$ have merely changed places. If $u > r$ and $v < s$, we have $w < ab$; this case can be transformed into the first one by the interchange of "B" and "$\bar{B}$". We therefore need only to consider the first case.

Using relations (7)–(8), we derive from (11):

$$u > a > r \; , \qquad v > b > s \; . \tag{13}$$

The first part of the first relation results from the first equation of (11), as follows. When we replace "r" by "u" in (11), we obtain $a = u$; if $r < u$, the value of a decreases. Similarly, we prove the second part of the inequality, and likewise the second relation in (13).

By a similar inference we derive from (10), using (7)–(8):

$$uv > w \; . \tag{14}$$

---

[1] *ThP*, p. 76.

For the proof we put "uv" for "rs" in (10); then we find that $w = uv$. Since $rs < uv$, the value of w is therefore smaller than that of uv.

Relations (13) and (14) read in probability notation:

$$P(C, A) > P(A) > P(\bar{C}, A) \ , \qquad P(C, B) > P(B) > P(\bar{C}, B) \ , \ (15)$$

$$P(C, A.B) > P(A.B) \ . \tag{16}$$

For the inverse probabilities we have similar relationships. Using the rule of the product in a form analogous to (2), we derive from (15)–(16):

$$P(A, C) > P(C) \ , \qquad P(B, C) > P(C) \ , \tag{17}$$

$$P(A.B, C) > P(C) \ . \tag{18}$$

These inverse probabilities can be computed as follows:

$$P(A, C) = \frac{u}{a} \cdot c \ , \qquad P(B, C) = \frac{v}{b} \cdot c \ , \tag{19}$$

$$P(A.B, C) = \frac{P(A.B.C)}{P(A.B)} = \frac{uv}{w} \cdot c \ . \tag{20}$$

From (13) we see that in each of these relations the factor preceding "c" is greater than 1. This means that the probability of the occurrence of the event C is raised when the event A, or the event B, or their combination, is observed. With these results, the statistical properties of the conjunctive fork are completely formulated.[2]

We have so far assumed that the time direction of the fork of figure 24 is known. Now let us assume that merely the time *order* is known; then we can as well draw the fork in the form of figure 25, replacing the "E" of figure 25 by "C". Can we now use conditions (5)–(8) to determine the time *direction* of the fork, that is, to distinguish between figure 24 and figure 25?

It is easily seen that this cannot be done directly. A common effect E, that is, the upper fork of figure 23, can very well satisfy the relations (5)–(8), with "E" in the place of "C". In fact, this is usually true. For instance, the spouting of the geysers may have the effect

---

[2]Sometimes the principle of the common cause is used in a weaker form, for which relations (5)–(6) are abandoned. It is then regarded as sufficient if the cause C satisfies the inequalities (1), (7)–(8), and (16). The inequalities (15) and (17)–(18) are then derivable. However, it is usually maintained that by the use of a more detailed description the cause C can be so specified as to satisfy (5)–(6) too. See the discussion of the between-relation in the remarks on p. 188 in §22, preceding the definition.

that two clouds are formed which merge into one large cloud. Then the occurrence of this large cloud is an effect E which satisfies (5)-(8).

However, there exists an indirect way of using relations (5)-(8) for a definition of time direction. These relations can hold for a common effect E only on the condition that there also exists a common cause C satisfying the same relations. That is, we then must have the double fork of figure 23, and not the single fork of figure 25. This is seen as follows. The inequality (1) is derivable from (5)-(8) just as well if these relations contain "E" in the place of "C". Therefore, if there exists a conjunctive fork with respect to a common effect E, the simultaneous occurrence of A and B is more probable than a mere chance coincidence. Consequently, if there were no common cause C, the common effect would establish a statistical dependence between A and B; and explanation would be given in terms of a "final cause". Referring to the discussion given in §18, we regard final causes as incompatible with the second law of thermodynamics and consider such forks impossible. In application to the present investigation this means: The principle of the common cause does not exclude, throughout, a statistical dependence with respect to a common effect; but it does exclude such dependence if there exists no common cause.

In contrast, it is quite possible that there is no common effect although there is a common cause satisfying (5)-(8). This result can be used for a definition of time direction, as follows.

If a fork ACB is conjunctive, we say that the fork is *closed* at C. If there is no event E on the other side of the fork which satisfies (5)-(8), we say that the fork is open on that side. We now define:

DEFINITION. In a conjunctive fork ACB which is open on one side, C is earlier than A or B.

This is a definition of time direction in terms of macrostatistics.[3]

---

[3] A definition of time direction based on forks of statistically connected events, for which only a common cause, not a common effect, can establish statistical correlations, was given in an earlier paper: see Hans Reichenbach, "Die Kausalstruktur der Welt und der Unterschied von Vergangenheit und Zukunft", *Sitzungsber. d. Bayer. Akad. d. Wiss.*, math.-naturwiss. Abt., 1925, pp. 133-175. The mathematical and logical relationships referring to probability implications given in that paper, however, are incorrect. It was written before my construction of a general theory of probability, and its results should be regarded as replaced by the relations developed in the present section. Furthermore, I believed that the direction of time was not derivable from the entropy principle. This conception I would now replace by the proof to be given presently, according to which such a derivation is possible if the entropy principle is supplemented by the hypothesis of the branch structure. The time definition of macrostatistics is of the same kind as that of microstatistics. For a further correction of views expressed in the paper mentioned, see §23, n. 1.

These results may be summarized in terms of the *principle of the common cause*, as follows: If coincidences of two events A and B occur more frequently than would correspond to their independent occurrence, that is, if the events satisfy relation (1), then there exists a common cause C for these events such that the fork ACB is conjunctive, that is, satisfies relations (5)–(8).

This principle means that a statistical dependence of simultaneous events requires an explanation in terms of a common cause. A common effect cannot be regarded as an explanation and thus need not exist, though it may exist and even may establish a statistical dependence.

There also exist, of course, forks which are open toward the past and thus are pointed toward the future, such as the fork of figure 25: but they never satisfy relations (5)–(8). For instance, when two trucks going in opposite directions along the highway approach each other, their drivers usually exchange greetings, sometimes by turning their headlights on and off. We have here a fork AEB, where E is the exchange of greetings, which is a common effect of the "coincidence" of the trucks, that is, of the events A and B. Here we have

$$P(E, A.B) > P(A) \cdot P(B) \ . \tag{21}$$

This means: From the effect E we can infer with high probability that two trucks met and passed. But relation (1) does not hold; instead, we have here

$$P(A.B) = P(A) \cdot P(B) \ . \tag{22}$$

This relation states that the common effect, the exchange of greetings, does not make the meeting of trucks more frequent. The fork is therefore not conjunctive.

Furthermore, there exist forks ACB in which the common cause does not establish a statistical dependence between A and B. For instance, let a die be thrown repeatedly; in the series of events thus resulting, the occurrence of face 6, which we shall call event A, has the probability $P(A) = 1/6$. Sometimes, a second die may be thrown with the same hand; if it shows face 6, we will speak of the event B. If A and B occur together, they go back to the same cause C, the throwing of the two dice. Although here relations (5)–(6) are satisfied, relation (7) is violated, because it is replaced by an equality. From (12) we then infer that $w = ab$; therefore (1) is replaced by an equality, and the fork is not conjunctive.

We shall now investigate the relationship between the common-cause principle and the second law of thermodynamics. At first sight, this principle seems to have no relation to Boltzmann's conceptions, because the occurrences to which it refers apparently do not resemble a mixing process. But when we recall the treatment of a mixing process by means of a probability lattice, presented in §12, we discover that this view is wrong. In fact, a lattice of mixture contains the very kind of relations to which the common-cause principle refers.

Let us assume that we have an ensemble of branch systems which includes two types of systems. In type T_A, one of the possible states is A; in type T_B, we find state B among the possible states. The successive states gone through by each system in the course of time are not strictly determined, if only for the reason that the system is not completely isolated from its environment; but each line of development can be regarded as a probability chain with aftereffect. Let us assume, furthermore, that sometimes, though not always, a system of type T_A and one of type T_B originate from only one state C of interaction, that is, that the two systems branch off from one causal root. It is this origin from the same causal event which leads to frequent combinations of states A and B, and is expressed by a statistical dependence.

Since we have here two different types of systems, we have two lattices x_{ki} and y_{ki}; let us imagine that the latter is directly on top of the former. The lattices are coupled by the event C in their first columns; whenever the lower lattice has a C in this column, there is a C directly above it in the upper lattice. Therefore, the same holds for $\bar{C}$, though the case $\bar{C}$ may represent a wider class of events than C—it merely means absence of C. Apart from this coupling, however, the lattices are independent. This condition can be written in the form

$$P\left(C^{k1},\ A^{ki}.B^{ki}\right)^k = P\left(C^{k1},\ A^{ki}\right)^k \cdot P\left(C^{k1},\ B^{ki}\right)^k\ ,$$

$$P\left(\bar{C}^{k1},\ A^{ki}.B^{ki}\right)^k = P\left(\bar{C}^{k1},\ A^{ki}\right)^k \cdot P\left(\bar{C}^{k1},\ B^{ki}\right)^k\ . \tag{23}$$

These relations represent conditions (5)–(6) transcribed into lattice form. Conditions (7)–(8) assume the form

$$P\left(C^{k1},\ A^{ki}\right)^k > P\left(\bar{C}^{k1},\ A^{ki}\right)^k\ ,\quad P\left(C^{k1},\ B^{ki}\right)^k > P\left(\bar{C}^{k1},\ B^{ki}\right)^k\ . \tag{24}$$

We can now derive relation (12), as before, and with it, relation (1). Therefore we have

$$P\left(A^{ki}.B^{ki}\right)^k > P\left(A^{ki}\right)^k \cdot P\left(B^{ki}\right)^k\ . \tag{25}$$

When we assume that the inequalities (24) are noticeable only if the distance from the cause C is not too large, that is, that relations (24) tend, with growing superscript i, to become equalities, we find

$$\lim_{i \to \infty} P\left(A^{ki} . B^{ki}\right)^k = P\left(A^{ki}\right)^k . P\left(B^{ki}\right)^k .\qquad (26)$$

Transcribed for relative probabilities, relations (25) and (26) assume the form:

$$P\left(A^{ki}, B^{ki}\right)^k > P\left(B^{ki}\right)^k , \quad P\left(B^{ki}, A^{ki}\right)^k > P\left(A^{ki}\right)^k ;\qquad (27)$$

$$\lim_{i \to \infty} P\left(A^{ki}, B^{ki}\right)^k = P\left(B^{ki}\right)^k , \quad \lim_{i \to \infty} P\left(B^{ki}, A^{ki}\right)^k = P\left(A^{ki}\right)^k .\qquad (28)$$

Relations (27)–(28) correspond to relations (12, 23-24) when we substitute "A" for "$\bar{B}$". One difference between them is that in (27)–(28) we do not use a superscript in the form "$k-1$". This difference results because in the lattice of §12 (12, 6) we have used alternatingly two different kinds of rows, whereas the present lattice is constructed as a superposition of two lattices, such that two events A and B which make up a pair belong to the same row.

There is a further difference. In the lattice of the events A and B, we do not assert an. analogue of relation (12, 22), which compares a vertical with a horizontal probability. In fact, when we continue the causal lines leading from C, or $\bar{C}$, to A or B, the latter events will usually not recur in these lines. We may compare the lattice of the events A and B to the entropy lattice (14, 1) and regard A and B as states of a very low entropy which do not recur in the history of a system before aeons of time have passed. Our common-cause lattice represents only the initial sections of the rows, in which the entropy increases; it does not extend into later parts of the rows. In fact, the lattice used for the common-cause principle may be regarded as a rudimentary form of a lattice of mixture, in which only the first columns are realized and the relations expressing mixture refer to vertical probabilities only.

Let us discuss from this viewpoint the double lattice controlled by relations (23)–(28). The first column contains "C" and "$\bar{C}$". In the rows beginning with "C" there is a higher vertical probability than in other rows for the occurrence of pairs "$A.B$", in a particular column i. Actually, "A" and "B" occur only in this column; in the idealization of the problem, we would assume that these events can also occur in later columns, though perhaps never twice in the same row. The influence of the arrangement of the first column shows in the statistical dependence of A and B in column i; this is expressed in relations (25) and (27). In the idealization of the problem, this statistical dependence of A and B would die down, as is formulated in (26) and (28).

The pairing of events in column i indicates a lattice of mixture. It refers to a mixing process in the mathematical sense, that is, a mixing process in which the elements are "shuffled", not by a physical mechanism, but by the relations of independence and lattice invariance. The statistical relations in the resulting lattice are the same as those holding for the mixture of molecules or the "mixture" of branch systems; they are visible, in the present lattice, through an internal dependence, or aftereffect, within a column, as discussed for the lattice of relation (12, 23). Thus the space ensemble here indicates mixture, even though probabilities referred to the time ensemble cannot be used. We will therefore regard the rudimentary lattice formulating the common-cause principle as subject to the same physical rules as the usual one.

On this condition the hypothesis of the branch structure applies to the present lattice, if this hypothesis is extended so as to include macrostatistics as well as microstatistics. The hypothesis tells us that if a state occurs more frequently in the space ensemble than corresponds to a certain standard, namely, to its probability in the time ensemble, there must have existed an interaction state in the past. For our rudimentary lattice, we have no time ensemble. We therefore replace this standard by a certain probability with reference to the space ensemble. This probability is constructed as follows.

If the two causal lines leading to A and B were not connected by their first elements, the probability of the joint occurrence would be given by

$$P(A) \cdot P(B) \tag{29}$$

both for time ensemble and space ensemble. Only the latter ensemble exists; so we take the value of (29) from the space ensemble. When we now regard the two lines as forming one system and the joint occurrence of A and B as a state of this system, we treat the value (29) as the probability of its time ensemble, that is, as the standard probability. Since in the space ensemble the conjunction $A.B$ is observed more frequently than corresponds to (29), it can be inferred that this deviation from the standard requires an interaction in the past. This means that the two lines must spring from an interaction process which included both of them.

It should be noted that these inferences correspond exactly to those we make for the usual lattices of mixture. For instance, in the lattice given by relations (12, 22–24) we have precisely these conditions. Here the probability of B in the time ensemble is countable in a row; but counting in any column, we find the same value for the space ensemble. The fact that, counting in the space ensemble, we find a frequency above the standard for a joint occurrence of B and $\bar{B}$, is expressed in relations (12, 23-24). This higher vertical probability is possible only because the rows of this lattice spring from an inter-action state in the past, a state in which the two gases started to interact, that is, entered into a mixing process.

The rule that a statistical dependence in a space ensemble requires an explanation in terms of a common cause, is thus seen to follow from the maxim that the observation of a state which is improbable in the history of a system requires explanation in terms of a preceding state of interaction. The rule follows when the expression "improbable in the history of a system" is replaced by reference to a standard prob-ability which can be measured in the space ensemble if the time ensemble does not exist, and which is identical with the probability in the time ensemble in case this ensemble does exist. The definition of time direction supplied by the principle of the common cause is thus identical with the rule that the observation of an ordered state in an isolated system requires an interaction state in the past, that is, that an ordered state is a postinteraction state. The simultaneous occur-rence of A and B is regarded as indicating order, because it takes place more frequently than is compatible with a chance coincidence.

If there were conjunctive forks open toward the past, this would mean that there were mixing processes in reverse, going from a state that occurs in the space ensemble more frequently than is warranted by the standard probability, toward a state of interaction. That such processes do not occur, that states of order are not preinteraction states, is asserted by the principle of the parallelism of entropy in-crease, formulated in assumption 5 of the hypothesis of the branch structure (§16). It was shown that this principle is the expression of a general statistical property of the universe, of its statistical isotropy. The common-cause principle thus reiterates the very principle which expresses the nucleus of the hypothesis of the branch structure.

● 20. *Entropy and Information*

Since the second law of thermodynamics leads to the existence of records of the past, and records store information, it is to be expected that there is a close relationship between entropy and information. This relation has recently become the subject of interesting investiga-tions in the theory of information, which has developed from studies concerning the transmission of information in telephone cables and the storage of information in computers. It has been shown that for proc-esses of this kind a measure of information can be defined which, in fact, has the properties of entropy and which has turned out to be an extremely valuable tool for the purpose of the information engineer. Since the concepts involved refer to probabilities ascertained in statistics of macroscopic objects, it is, therefore, necessary to define an entropy measure for macrostatistics, that is, the concept of macro-entropy.

The term *information* refers to a *probability disjunction*, or *probability situation*. A disjunction of this kind is given by a complete and exclusive disjunction of events $B_1 \vee \ldots \vee B_m$ for which a set of probabilities $p_1 \ldots p_m$ is known. It follows that the p_i satisfy the condition

$$\sum_{i=1}^{m} p_i = 1 . \tag{1}$$

Let us ask how the information conveyed by such a disjunction can be measured.

Assume first that the disjunction has only two terms and that we thus have only the probabilities p and $1-p$. We shall say that the disjunction conveys much information to us if by using it we are able to make a large number of predictions which come true. We have the choice of the prediction; therefore we shall predict either B, or $\bar{B}$, depending on whether p, or $1-p$, is greater than $1/2$. The measure of information then would simply be given by p, or by $1-p$, whichever is the larger. The maximum of information would be equal to 1; the minimum, equal to $1/2$. Information of this kind will be called information *about* the event, or *intensional information*. It increases with the percentage of true predictions and thus expresses the *degree of predictability* associated with the disjunction.

There exists, however, a second kind of information. When we make a prediction which turns out to be false, we learn something from the observation; that is, the event imparts information to us. No such information is given when the prediction is true, because the event then merely tells us what we expected. This second kind of information will be called information *from* the event, or *extensional information*. Its measure increases with the number of false predictions. Extensional information is the inverse of intensional information. The term "information" in its usual sense means "intensional information"; but sometimes extensional information is convenient for practical purposes. Obviously, if one kind of information is defined, the other kind is also defined; the two forms represent merely different modes of speech. It is a difference like the one between hot and cold, or between large and small. In the following, we shall use the word *information* in the sense of intensional information; when we refer to extensional information, this will be explicitly specified.

The value p, or $1-p$, would be a possible measure of information for a disjunction of two terms; but it is inappropriate for larger disjunctions. Assume that we have the three probabilities p_1, p_2, p_3, and assume that $p_1 > p_2 > p_3$. We shall then predict the event B_1, and shall be right in the percentage p_1 of all cases. If we now use p_1 as the measure of information, the difference between the values p_2 and p_3 is not expressed. Let us therefore look for a different measure. We shall permit a double prediction, the second part in a conditional form: first, we predict B_1, and for the case that B_1 does not occur, we predict B_2. This prediction comes true in $p_1 + p_2$ cases. If we use $p_1 + p_2$ as a measure of information, we have thus expressed the fact that p_1 as well as p_2 is larger than p_3; but we have not expressed the difference between p_1 and p_2. In fact, predicting first B_2 and, in case it does not occur, B_1, leads to the same score.

We will therefore revise the computation of the score as follows. We shall attach a higher weight w_1 to the prediction if the first part comes true; and a lower weight w_2 if the second part comes true. We may imagine a man making the double prediction repeatedly: every time the first part of his prediction comes true, the value w_1 is scored for him; and every time the second part comes true, the value w_2 is credited to him. The sum total of the credited w_i is then divided by the number n of observed cases. Using this score, we are interested in predicting B_1 for the first place, and B_2 in the second; we then shall arrive at the score

$$p_1 w_1 + p_2 w_2 \ . \tag{2}$$

If someone were to predict B_2 first, he would arrive at a score in which the values p_1 and p_2 have changed places and which is thus smaller.

Once we have introduced weights, we may as well allow for a triple prediction, by adding to the previous one the conditional form: if neither B_1 nor B_2 comes true, B_3 will occur. This triple prediction, of course, will always come true; but by assigning to the third place the smallest weight w_3, we construct a score

$$p_1 w_1 + p_2 w_2 + p_3 w_3 \ ;$$

$$p_1 > p_2 > p_3 \ , \qquad w_1 > w_2 > w_3 \ , \tag{3}$$

in which the differences in the p_i are expressed and which compels us to order our predictions in relation to the degrees of the probabilities. The addition of the third term has the technical advantage that the measure of information is referred to the whole disjunction and not only to a part of it. And, as before, information is thus a measure of the degree of predictability associated with the disjunction.

The measure (3) can be extended to disjunctions of any length m, including the case $m = 2$. Here, obviously, the measure (3) increases directly with the higher of the two probabilities p and $1 - p$ and is thus identical with the measure discussed previously as far as the order of possible scores is concerned, though it may be metrically different. In the investigations concerning information theory[1] it has been shown that it is convenient to use for the w_i the special form

$$w_i = \log p_i \ . \tag{4}$$

This form has the advantage that the value w_i automatically increases with p_i; we thus need not state explicitly the conditions expressed in the inequalities in (3). Since the logarithm of a fraction smaller than 1 is negative, it is advisable to add to the expression (3) a positive constant so as to make the total score positive. We thus arrive at the following definition of the measure of intensional information:

$$H(p) = \sum_{i=1}^{m} p_i \ \log p_i + \log m \ . \tag{5}$$

It can be shown that for all possible values of the p_i, $H \geq 0$. The function H reaches its lowest value 0 if all the p_i are equal and thus $p_i = 1/m$. It reaches its maximum if one of the p_i is equal to 1 and all the others equal 0. In this case, the expression under the sum is equal to 0 for every p_i; and we have $H = \log m$ for the maximum of H. This is, in fact, the situation of highest predictability, that is, of highest information about the event. The minimum $H = 0$ corresponds to the indifference situation in which no B_i deserves preference over the others. We saw above that for $m = 2$ this is, in fact, the lowest degree of predictability. In figure 26, the form of $H(p)$ for $m = 2$ is represented as a function of p_1.

If we have two probability disjunctions $p_1 \ldots p_m$ and $p_1' \ldots p_m'$, we can ask for the combined information which they give us. This means, we ask for the information concerning possible combinations of the events to which the probabilities refer. Let us assume that the events are independent; then we can multiply the probabilities, and have

$$H(p \cdot p') = \sum_{i=1}^{m} \sum_{k=1}^{m} p_i \, p_k' \ \log p_i \, p_k' + \log m^2$$

$$= \sum_{i=1}^{m} p_i \ \log p_i \sum_{k=1}^{m} p_k' + \sum_{k=1}^{m} p_k' \ \log p_k' \sum_{i=1}^{m} p_i + 2 \log m \ . \tag{6}$$

---

[1]See p. 19 of Claude E. Shannon and Warren Weaver, *The Mathematical Theory of Communication* (Urbana, Ill., Univ. Ill. Press, 1949). The func-

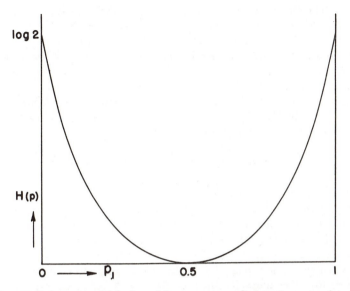

Fig. 26. Information $H(p)$ plotted for a two-term disjunction associated with the probabilities p_1 and $p_2 = 1 - p_1$, according to formula (5). The minimum value $H = 0$ is assumed for $p_1 = p_2 = 1/2$; the maximum value $H = \log 2$ for $p_1 = 0$ and for $p_1 = 1$.

Using (1), we thus find

$$H(p \cdot p') = H(p) + H(p') . \tag{7}$$

This means, for independent events, information is additive.

It is sometimes convenient to introduce a generalized measure of information. Suppose we believe that a disjunction $B_1 \vee \ldots \vee B_m$ is associated with the set of probabilities $q_1 \ldots q_m$, whereas it is actually controlled by the probabilities $p_1 \ldots p_m$. When we now make predictions, we shall write down for every observed event B_i the score $\log q_i$, according to our knowledge of the situation. As usual, the sum total of credited amounts is divided by the number n of observed cases. Since the frequencies of the occurrence of the B_i are actually controlled by the p_i, we thus arrive at the score

$$H(p ; q) = \sum_{i=1}^{m} p_i \, \log q_i + \log m . \tag{8}$$

tion H which Shannon and Weaver use is, however, the negative of the function given in equation (5) of the current section, and is thus extensional information; furthermore, the constant $\log m$ is omitted. See also Norbert Wiener, *Cybernetics* (Cambridge, Mass., Technology Press, 1948), chap. iii.

We call this the information by $q_1 \ldots q_m$ in the situation $p_1 \ldots p_m$. Since our observational statistics will soon show us that the q_i do not apply to this situation, we shall eventually replace them by the p_i, which we may take from the statistics of observations. We will therefore also call (8) the *antecedent information* by the $q_1 \ldots q_m$ in the situation $p_1 \ldots p_m$.

When we now proceed to using the correct weights $\log p_i$, we arrive at the score (5). Therefore we have improved our score by the amount resulting when we subtract (8) from (5), that is, by

$$H(p/q) = \sum_{i=1}^{m} p_i \log p_i - \sum_{i=1}^{m} p_i \log q_i \; . \tag{9}$$

We call this expression *differential information*, the information by $p_1 \ldots p_m$ with respect to $q_1 \ldots q_m$. The constant "$\log m$" drops out here. It can be shown that (9) is always ≥ 0. It reaches its minimum value 0 if $p_i = q_i$. If q_1 is the smallest value among the q_i, (9) reaches its maximum for $p_1 = 1$; this maximum is equal to $-\log q_1$.

When all the q_i are equal to $1/m$, the expression (9) becomes identical with (5). The plain information (5) may therefore be regarded as differential information with respect to the indifference situation; that is, we have

$$H(p) = H\left(p \Big/ \frac{1}{m}\right) \; . \tag{10}$$

This shows that we actually always measure informational differences, and that therefore the general form (9) is the most appropriate.

Formula (9) is illustrated in figure 27 for $m=2$. We put here

$$H(p/q) = H_1 - H_2 \; ,$$

$$\tag{11}$$

$$H_1 = p \, \log p + (1-p) \, \log (1-p) \; , \qquad H_2 = p \, \log q + (1-p) \, \log (1-q) \; .$$

The two curves H_1 and H_2, which are throughout ≥ 0, are drawn in the lower part of the diagram as functions of p. H_2 is a straight line tangent to H_1 at the point $p=q$. It is easily seen that their difference $H_1 - H_2$ is throughout greater than or equal to 0. This curve, which is $H(p/q)$, is drawn in the upper part of the diagram. For instance, the ordinate $X_2 X_1$ is equal to the distance $Y_2 Y_1$. Formula (5) represents the special case resulting for the horizontal tangent at the lowest point of the curve H_1. For $m>2$ the diagram is essentially the same; H_2 represents here the plane tangent to the surface H_1.

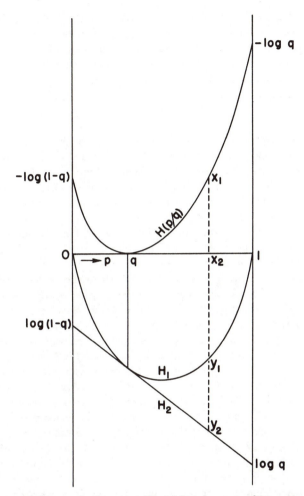

Fig. 27. Differential information $H(p/q)$ for a two-term disjunction associated with the probabilities p and $1-p$; q and $1-q$ are the antecedent probabilities. See formulas (9) and (11). The minimum value $H = 0$ is assumed for $p = q$.

By analogy with (7), we can prove that $H(p/q)$ is additive, if the probabilities can be multiplied. The proof is given by analogy with (6), when we write (9) for combinations $p_i p_k'$ and $q_i q_k'$, and interchange summations suitably. We thus find

$$H(p \cdot p'/q \cdot q') = H(p/q) + H(p'/q') .\qquad(12)$$

This relation shows that differential information is additive for independent events.

To those who are acquainted with the results of the kinetic theory of gases, formula (5) looks familiar. In fact, we derived this formula in (8, 15), applying Boltzmann's ideas to the problem of the distribution of gas molecules in a container. We saw there that H is merely a negative entropy, being the negative of the specific entropy S computed in (8, 11). That is, we have

$$H = -S \ . \tag{13}$$

Vice versa, we may therefore call entropy inverse information. In fact, entropy can be regarded as extensional information. It is in view of this relationship that the form (4) has been selected for the weights determining the informational score. To examine the connection between entropy and information more closely, let us consider the following illustration.

Suppose the gas is in a state of partial order, that is, not in a state of equilibrium. The entropy of the gas has then a certain medium value. The distribution is known when we know for every cell how many molecules are contained in it, that is, when the probabilities p_i introduced in (8, 10) are known. The molecules may be regarded as being numbered individually, although the order of these numbers, or proper names of molecules, is in no relation to the existing arrangement.

Now imagine a man who knows the distribution and who is asked to tell, without looking into the container, in which cell the molecule number 1 is at present. His information then consists in the knowledge of the probability disjunction $p_1 \ldots p_m$, and he will make the conditional prediction described. If the prediction is repeated for molecule number 2, number 3, and so on, his score will be measured by (5). Consequently, if the p_i are very unequal and the entropy is low, he will make a high score, that is, his information is high. Conversely, if the p_i are practically equal to one another and the entropy is high, he will make a low score; his information is low. Entropy increases with ignorance about the event predicted and is in inverse ratio to intensional information. However, if we are ignorant *about* the event, we shall learn much *from* the event; entropy therefore measures information *from* the event, or extensional information.

Once this relation between entropy and information is recognized, however, it suggests a second interpretation of the measure of information. The identity of formulas (5) and (8, 15) shows that relation (5) can be given a meaning different from the one we used in its derivation. Suppose that the probability situation $p_1 \ldots p_m$ is repeated n times, such that each time an event is produced. There results a sequence of n events among which the frequencies of $B_1 \ldots B_m$ are

controlled by the p_i; such is the information given to us by the probability disjunction $p_1 \ldots p_m$. But we do not know in what order the n
events will occur. There exists a class of arrangements of the n elements of the sequence such that each of these arrangements satisfies
the frequency conditions imposed upon the sequence by the $p_1 \ldots p_m$.
The number N_D of arrangements within this class is found by permuting
the elements, according to Bernoulli's theorem; this number N_D, therefore, is measured by formula (8,5), and its logarithm, given by (8,7),
is identical with the negative value of the summation term in (5). If the
measure of information increases, the number N_D goes down, whereas
if negative information, or entropy, increases, the number N_D goes up.
Entropy is therefore a measure of the extension of the class of possible arrangements; in the form of absolute entropy, it is an absolute
measure, and in the form of relative entropy, a relative measure.

The probability disjunction informs us that an arrangement of this
class will occur, but does not tell us which; therefore, it tells us much
when the class is small, and tells us little when the class is large,
leaving many possibilities open. The intension of the information thus
is high when the extension of the class of arrangements is small. When
we define information as a negative entropy, we merely follow the law
of inverse correlation between intension and extension, familiar from
traditional logic. Formula (5) appears as a reasonable measure of the
intensional information given by a probability disjunction, because its
negative value measures the extension of possible arrangements compatible with this disjunction; and this negative value, the entropy, is
thus rightly regarded as extensional information.

This relation between entropy and information can always be applied, even if the entropy does not have the simple form (8, 11). Entropy
measures the extension of the same-order class, as was explained at
the end of §9. A low entropy always determines a small class of arrangements; and a high entropy, a large one. Using the law of intension
and extension, we shall therefore assign high intensional information
to a low-entropy state. The corresponding probability disjunction, that
is, the set of probabilities having this measure of information, is given
not by the probability which measures the entropy, that is, not by the
probability W of (8,11), but by the frequencies with which molecules
occur in the different cells of the space considered, that is, by the p_i
of (8,11).

The general measure of differential information (9) can be given an
interpretation similar to that of (5). Formula (9) results when the cells
in the gas space do not have equal antecedent probabilities, but when
these antecedent probabilities are equal to q_i. Formula (8,4) is then
replaced by the relation

$$W = N_D \cdot q_1^{n_1} \cdots q_m^{n_m} . \tag{14}$$

Applying (8, 7) to the n_i of this formula and using (8, 10), we easily derive

$$S = \frac{1}{n} \log W = - \sum_{i=1}^{m} p_i \log p_i + \sum_{i=1}^{m} p_i \log q_i . \tag{15}$$

This is the negative value of (9). The measure of differential information corresponds to a generalized form of entropy, for which the cells of the parameter space have unequal antecedent probabilities. Although such forms of entropy are not used for microstatistics, they may play an important part in macrostatistics.

The distinction between relative and absolute entropy, discussed in §8, finds its analogy in a difference between relative and absolute information. The measure (5) of information is a relative measure, because (5) is identical with the negative value of (8, 11). Omitting the constant "$\log m$" in (5) corresponds to omitting the constant "$-\log m$" in (8, 11), and thus to using an absolute entropy, given directly by the logarithm of the number of arrangements, that is, by (8, 7). This absolute form of extensional information is used by C. Shannon.[2] It is a consequence of the omission of the constant that, for an equiprobable disjunction, the entropy increases with the length m of the disjunction. The corresponding intensional information would then decrease. In contrast, if the form (5) is used, the intensional information is the same, namely, equal to 0, for any length m of an equiprobable disjunction. The maximum of the intensional information, resulting when one of the probabilities is equal to 1, changes in an opposite way: it is the same for all m when the constant is omitted, but increases with m if the constant is used. There is no a priori reason to prefer one form of definition to the other. Decrease of intensional information with growing m in equiprobable disjunctions, associated with absolute information, appears adequate if the disjunction "$B_1 \vee \ldots \vee B_m$" is made longer by adding terms—for instance, by adding cells of equal size in the space considered. However, when the larger disjunction results from using smaller cells in the same total space, it appears adequate to regard the information as remaining unchanged for equiprobable cells, a condition which is satisfied for relative information. It is then plausible that the maximum intensional information goes up.

The remark may be added that an analogue of Gibbs's paradox (see §8) cannot be constructed within information theory. Formulas (7) and

[2] See Shannon and Weaver, *op. cit.*, pp. 18–22.

(12), in which information is additive, correspond to the computation of the entropy for two containers, placed side by side, on the condition that no window between the containers is opened. Under these circumstances, the entropy is additive, too. The physical process of opening the window, which leads to a mixture of the gases and, for equal gases, to Gibbs's paradox, has no analogue in information theory. For this reason, an absolute-normalized measure of information can be carried through consistently. However, for the purpose of the present investigation, the relative-normalized information (5) appears more convenient.

With the definitions developed in this section, a typical explication problem has been solved: the problem of replacing the vague term "information" of everyday language by a precise one which matches the usage of the term. When we speak here of matching the usage, we know very well that we can only expect a qualitative correspondence. It was pointed out at the end of §2 that more cannot be attained for the very reason that the concept of everyday language, or explicandum, is vague and we can never be sure that we have found its exact meaning. The precise concept, or explicans, should be regarded as a proposal for a new usage in which the vagueness of the explicandum is eliminated, while its meaning is adhered to as closely as possible.

It is a remarkable achievement that such an explication can be given for a concept as abstract and indefinite as that of information. I could very well imagine that many an old-style philosopher would have insisted that it is impossible to measure information, that this concept is one of many unmeasurables which a mystically inclined philosophy is so proud to possess. Mathematicians, however, not inhibited by philosophical scruples, had the courage to go to work and construct a measure of information which satisfies the needs of the communication engineer and which, at the same time, supplies the philosopher with an instrument of logical research.

The connection between information theory and statistical mechanics is a major result of these investigations; yet this very connection also confronts us with a new and significant question. The inverse relation between entropy and information has been rightly emphasized by those mathematicians to whom we owe the inception and elaboration of information theory, and among whom the names of Von Neumann, Wiener, and Shannon are prominent. Yet what has been proved is merely that the entropy of gas theory has the character of negative information. Can we conclude that, vice versa, the information of information theory has the character of negative entropy? That is, can the macroentropy, defined as the negative of intensional information (5) and thus as extensional information, be regarded as an indicator of time direction?

The presentations of information theory, it is true, include remarks that the entropy therein defined is closely related to the second law of thermodynamics. But the proof that this is true and that the entropy of information theory does define a time direction, can only be given by the use of the results derived for recording processes from the hypothesis of the branch structure. In the following section this proof will be given.

• 21. *The Time Direction of Information and the Theory of Registering Instruments*

In order to show that information processes indicate a direction of time, we will show that such processes represent an application of the principle of the common cause.

Let us first show that, vice versa, the principle of the common cause is translatable into information theory. We use the notation:

$$P(A \cdot B) = p_1 \ , \quad P(A \cdot \bar{B}) = p_2 \ , \quad P(\bar{A} \cdot B) = p_3 \ , \quad P(\bar{A} \cdot \bar{B}) = p_4 \ ; \tag{1}$$

$$P(A) \cdot P(B) = q_1, \ P(A) \cdot P(\bar{B}) = q_2, \ P(\bar{A}) \cdot P(B) = q_3, \ P(\bar{A}) \cdot P(\bar{B}) = q_4 \ .$$

Relation (19, 1), which asserts that the frequency of the conjunction $A \cdot B$ is higher than would correspond to chance coincidences, can now be expressed as follows, when we use the differential information formula (20, 9) for $m = 4$:

$$H(p/q) = \sum_{i=1}^{m} p_i \ \log p_i - \sum_{i=1}^{m} p_i \ \log q_i > 0 \ . \tag{2}$$

This means: the deviation of the observed frequencies from a chance distribution presents us with information greater than 0. Note that this formulation is given in a neutral way, in which the conjunction $A \cdot B$ is put on a par with the three other possible conjunctions.

Using the relation between information and entropy formulated in (20, 13) we can also say that (2) measures the negative macroentropy of the process represented by the coincidences of A and B. A high amount of information means that the macroentropy is low. Therefore, unusual coincidences indicate low entropy.

We know that a low-entropy state requires an interaction in the past for its explanation; and we saw that the same holds for relation (19, 1).

We conclude that the occurrence of information requires past inter-action, that is, that information can be supplied only by postinteraction states.

This can be concluded, at least, when information consists in a deviation of the frequency of occurrences from a chance distribution. In §18 we studied natural records, such as footprints in the sand and fossils in the ground. They are all of this kind: chance would not produce those phenomena, which we interpret rightly as the product of past causes. It is not different with artificial records. When we write on paper, or record music on phonograph disks, or store information in the "memory" of an electronic computer, there result specific arrange-ments of elements, like ink marks on paper, grooves in wax disks, electric charges in little condenser elements, which are too highly ordered to be interpreted as a product of chance. For this very reason, they are records and supply information. All information is a record of the past. It may also inform us about the future, but this is possible only when we use the report of the past as a basis for inductive in-ferences. We need not refer to such further usage of information in this context, in which we ask for the source of information. In any case, its source is always in the past.

To study a suitable example, let us investigate the logical mecha-nism of registering instruments. They are used to register observables which vary irregularly and cannot be predicted. For instance, a baro-graph registers measurements of atmospheric pressure, the values of which vary irregularly, on a rotating drum. The zigzag line drawn by the pen of the instrument on a sheet of paper contains the record of these irregular variations.

It would be impossible for us ever to predict the varying values of the atmospheric pressure, but it is easy to record them. We face here a striking difference between past and future, which was mentioned earlier, in §2. For physical quantities of irregular variation, *prediction* is technically impossible, whereas *postdiction* is easily achieved by the use of registering instruments. The difference can be explained in terms of the relations holding for conjunctive forks; and thus statement 5, §2, can now be explained.

In a registering instrument, the physical quantity measured—for example, the atmospheric pressure—is the cause C which produces the effect A, the reading registered on the paper. It has many other effects B, too; for instance, the atmospheric pressure can cause a storm which fells trees. The reading A is merely a partial effect; yet it is sufficient to permit us to infer the total cause, the value of the physical quantity. The *inference from the partial effect to the total cause* is typical for all forms of recording processes.[1]

This explains why registering instruments are always constructed by the use of a cause which is common to the record and the other effects of the measured quantity, and thus by means of a fork which is pointed toward the past. If we could construct a registering instrument of the future, we would have to use a fork pointed toward the future; but in such a fork the corresponding inferences could not be made. We cannot make an inference from the partial cause A to the total effect E, except for a particular form of arrangement: we can do so in a double fork such as is indicated in figure 23 (see §19), that is, if there is a common cause relating A and B. This was referred to in §2 and §20. Let us now illustrate it in more detail. From a low atmospheric pressure measured in one place we can infer with some probability that it will rain over a larger area. This inference has the form of the schema given in figure 28. Here A is the reading on the barograph; C, the atmospheric pressure at the place of the instrument; B, some other effects of this local pressure; C_1, the atmospheric pressure over a wider area; B_1, other effects of this pressure; and E, the rain. The fork ACB is open toward the future; but the fork CEB_1 is closed on both sides, and therefore the partial cause C can be used for an inference toward the total effect E. Such an inference toward the future is made possible by a detour through the past, as it were.[2]

Since the fork ACB of figure 28 is open on one side, it must be pointed toward the past. The record A itself, the ink marks on paper,

---

[1][The terms "total cause", "total effect", "partial cause", and "partial effect" should be construed in reference to causal forks as explained in §19. In particular, the term "total cause" is not meant to refer to the sum total of events which have contributed causally to a given effect, but rather to the totality of causes appearing in a causal fork. Thus "total cause" refers to a causal fork closed toward the past; it denotes the common cause of multiple effects, each of which is a "partial effect". Similarly, "total effect" and "partial cause" have reference to causal forks closed toward the future.

Admittedly, causal forks and diagrams of them are schematizations which constitute a simplification of actual situations. Yet it would be beside the point to object to the use of the term "total cause" on the ground that in reality no single event is ever the sole cause of any effect. On the contrary, when the author speaks, for instance, of a total cause, that event *is* a total cause *within the schema in question*, and that is all that is meant.—M. R.]

[2]This discussion of registering processes and the inference from partial effect to total cause, which has no parallel in an inference from partial cause to total effect, was given in my earlier paper mentioned in a note to §19: see Hans Reichenbach, "Die Kausalstruktur der Welt und der Unterschied von Vergangenheit und Zukunft," *Sitzungsber. d. Bayer. Akad. d. Wiss.*, math.-naturwiss. Abt., 1925, pp. 133-175.

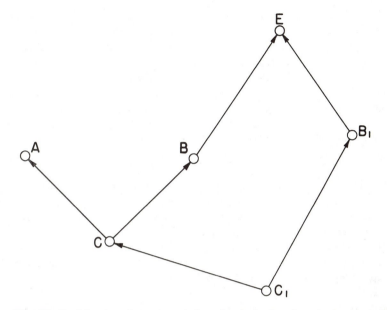

Fig. 28. Combination of an open fork and a fork closed on both sides. The observation of the record A of the event C makes possible the prediction of the later event E.

does not contribute to the rain. But it permits us to make an inference toward the cause C, although it is only a partial effect of C.

In contrast, if a fork is pointed toward the future and open toward the past, we have to know the occurrence of both A and B when we wish to predict the common effect E with a high probability. This may be illustrated by an example used in §19: When we see two trucks approaching each other, we can predict with high probability that the drivers will exchange greetings by light signals; but we cannot make this prediction when we see only one truck coming. It is for similar reasons that predictions usually require a large number of observational data. In general we have here multiple forks, in which a number of simultaneous events $A_1 \ldots A_n$ determine a common effect E. The results derived for binary forks can, of course, be easily extended to multiple forks. For instance, it was mentioned in §2 that a prediction of the atmospheric pressure requires data taken from a wide area; this is necessary because for meteorological occurrences we have no forks closed in the past. The same applies to predictions of political occurrences and of war. Here we have to make an inference from the total cause to the effect; this is why meteorological and political predictions are so difficult to make.

The mathematical treatment of the information conveyed by a registering instrument can be given as follows. To simplify the problem, assume that the instrument does not indicate continuously, but is capable only of r different indications $A_1 \ldots A_r$. Correspondingly, the observed quantity—for example, the air pressure—is classified into r possible states $B_1 \ldots B_r$. We may identify the value of the observable during the act of registration and the value shortly afterward, so as to be able to interpret the B_i as events coinciding with the A_i, in the sense of the principle of the common cause. We let

$$P(A_i) = a_i , \quad P(B_i) = b_i , \quad P(A_i . B_k) = p_{ik} . \tag{3}$$

If the instrument is perfect, that is, always indicates correctly, we observe only pairs $A_i.B_i$; we therefore have $p_{ik} = 0$ for $i \neq k$, and $p_{ii} = a_i = b_i$. The values a_i, b_i, and p_{ii} can be ascertained from statistics compiled for the instrument over a long period of time, in which we count the relative frequency of states A_i, B_i, and A_iB_i, for every i. The perfect registering instrument can therefore be characterized by the following relations, constructed by analogy with (1):

$$p_{ik} = 0 \text{ for } i \neq k , \qquad p_{ii} = a_i = b_i ;$$

$$q_{ik} = a_i b_k , \qquad\qquad q_{ii} = a_i^2 . \tag{4}$$

When we wish to find the differential information defined by (2), we may use double summations, a method of writing which is equivalent to counting all the values p_{ik} in a series of $m = r^2$ terms. We thus find

$$H(p/q) = \sum_{i=1}^{r} \sum_{k=1}^{r} p_{ik} \log p_{ik} - \sum_{i=1}^{r} \sum_{k=1}^{r} p_{ik} \log q_{ik}$$

$$= \sum_{i=1}^{r} a_i \log a_i - \sum_{i=1}^{r} a_i \log a_i^2 \tag{5}$$

$$= -\sum_{i} a_i \log a_i = H^*(a) ,$$

$$H^*(a) = -[H(a) - \log r] . \tag{6}$$

The last line in (5) follows when we put "$2 \log a_i$" for "$\log a_i^2$". Formula (5) means that the *intensional* information H conveyed by the p_{ik} relative to the q_{ik} is equal to the *extensional* information H^* conveyed by the a_i, if the latter information is measured absolutely, that is, without the additive constant "log r" contained in $H(a)$. This result, which holds only for a perfect registering instrument, defined by relations (4), explains the use of the form "$H^*(a)$" as a measure of information in such recording arrangements as those studied by C. Shannon.[3] However, the form "$H^*(a)$" does not show openly that it measures the degree to which coincidences of A_i and B_i are more frequent than would correspond to their independent occurrence.

If the recording process is repeated n times we may add up the individual amounts H given by (5), according to (20, 12), and thus have

$$H_n = n \cdot H \tag{7}$$

for the total information H_n conveyed by n successive acts of recording. For this result the content of the individual recorded item does not matter, since information is defined in a neutral way for the whole disjunction. This additivity is restricted to independent values of the observable; but this condition is usually satisfied. The increase in information with growing number of recorded data thus finds a natural expression.

In statistical interpretation, this additivity of information means that the larger the number of recorded items is, the lower is the probability that the items are a product of chance, and the higher the probability that they represent a record of the past. Strangely enough, this leads to the consequence that growing order is an indication of positive time, contrary to the entropy rule, according to which positive time tends to produce disorder. But this apparent contradiction is easily resolved. The order of the registered items does not represent a succession of states in an isolated system, but results from the space ensemble of individual interaction states, each of which produces a specific recorded item. And order in the space ensemble indicates a time direction perpendicular to the ensemble. We thus have here a time sequence the order, or information content, of which grows with positive time, as expressed in (7).

It is a consequence of this fact that, with increasing information, the manifestation of time direction becomes more and more pronounced. This peculiar feature of registering instruments is shown very clearly when we describe the registering process in opposite time. In language L_2 (see §16), the occurrences at a barograph would be described as follows. The drum on which a zigzag curve is drawn rotates backward, while the lever ending in the pen moves up and down under the effect of atmospheric pressure. It is controlled by those states of the air

---

[3]See Claude E. Shannon and Warren Weaver, *The Mathematical Theory of Communication* (Urbana, Ill., Univ. Ill. Press, 1949), pp. 19–21. Shannon writes it as "$H(a)$".

pressure which in L_1 follow immediately the position of the pen; since here the air pressure is practically the same as in the states preceding in L_1 the position of the needle, we have here nothing unusual. However, we observe a miraculous coincidence in the change of the air pressure and the shape of the zigzag curve: the pen is always exactly at that spot of the paper which is marked by the curve. The curve, therefore, predicts the air pressure. Finally, the ink flows back from the paper into the pen, and the ink marks disappear.

The latter process is unusual because the flow of ink toward the paper and the drying of the ink are irreversible processes. But it is not this fact alone which makes the whole occurrence unusual. It is the permanent coincidence of the curve and the position of the pen which we must study. We could, in fact, replace the writing process of the pen by some reversible process; then the permanent coincidence of curve and position of the pen would still remain incredible.

A registering instrument using a reversible process to record the observations can be constructed as follows, when we use language L_1 for the description. On a moving band, small posts are mounted, around which little cogwheels can rotate (fig. 29). The aneroid, which is mounted outside the band, controls a pointed lever which can move up and down and is in this way raised or lowered indirectly by the air pressure. When the pointed lever hits one of the cogwheels, this wheel starts spinning, pushed by the collision, the impact stemming from the motion of the band. The wheel retains its motion indefinitely, if we make friction as small as possible. While the band moves slowly to the left, one wheel on each post is rotating all the time, and the value of the atmospheric pressure is thus recorded in the height of the spinning wheel. In other words, the irregular variations of the atmospheric pressure are registered in the state of order of the wheels, and the distribution of the spinning wheels assumes the same function as the zigzag curve of the usual barograph.

This distribution is stored information. It can be measured by (5), if we understand by $A_1 \ldots A_r$ the different positions of the spinning wheel on a post and ascertain the a_i by counting the relative frequency of the states $A_1 \ldots A_r$ if the instrument is used over a longer time period. If we regard the combination of the spinning wheel and the corresponding air pressure as one physical system, the macroentropy of that system is given by the negative value of (5), up to an additive constant. Each act of registration represents such a system. We can also regard a number of such systems as forming one comprehensive system. Its macroentropy is given by the negative of the sum of the amounts H computed from (5), that is, by $-H_n$ as determined in (7). The

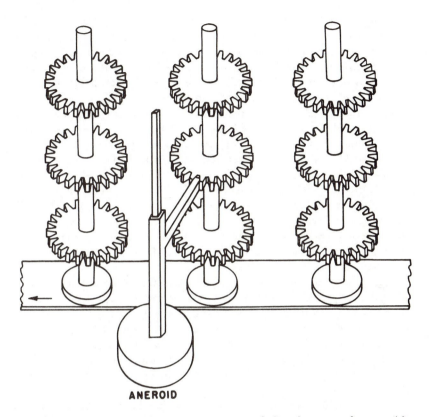

ANEROID

Fig. 29. A registering barometer constructed by the use of reversible processes. The pointed lever is raised or lowered by the aneroid. The columns move with the band, and the wheel touching the pointed lever is pushed and starts rotating. The air pressure is registered in the distribution of the spinning wheels. Interpreted in reverse time, the registering process does not contradict physical laws, but it leads to improbable coincidences.

very low entropy of this system shows that it is highly ordered, which is to say, it conveys much information. We thus have here a mechanism constructed entirely by means of reversible macroprocesses which indicates time direction by a low value of its macroentropy, in the same way that a low value of microentropy indicates preceding interaction. The order of the spinning wheels reveals time direction by the principle by which we infer that a burning cigarette must have been lighted shortly before.

The existence of a direction in this mechanism is made particularly obvious when we describe the processes in language L_2. Speaking the

language of negative time direction, we observe the following coincidences. The band moves from left to right, carrying the posts, on each
of which one wheel is spinning. The heights of these spinning wheels
vary irregularly. When the post passes the lever of the aneroid, this
lever is always at the correct height so as to hit the spinning wheel,
the motion of which is then stopped by the lever.

This stopping of the spinning motion is the reverse of the process
which we call in L_1 an impact imparted to the wheel. Since the impact
is a mechanical process, it may be regarded as reversible, if we ignore
friction and consider the impact perfectly elastic. The description of
the process in L_2, therefore, does not contradict physical laws. However, in L_2 we continually observe a coincidence between the position
of the lever and the position of the spinning wheel that arrives. It is
this series of improbable coincidences which requires an explanation.
This explanation could be given in L_2 only by reference to the common
effect: it would seem to be the purpose of these occurrences to stop
the spinning motion of the wheels.

In contrast, the explanation in L_1 would refer to the common cause.
That the height of the spinning wheel, soon after it was hit by the
lever, still corresponds to the height of the lever, is a coincidence
explained by the fact that the lever caused the spinning of the wheel.
Although this registering instrument records by means of reversible
processes, it leads to a definition of time direction in terms of the
principle of the common cause and thus defines time direction in terms
of macrostatistics.

These considerations show why registering instruments can record
only the past, and not the future, and why information defines the same
time direction as growing entropy. The registering of information reveals the unidirectional nature of time, because only a past cause can
produce a correspondence between signs and physical data, and because an increasing amount of information represents the order of a
space ensemble which calls for explanation in terms of a past cause.
The time direction of information is the direction imprinted upon the
universe by its statistical isotropy.

● 22. A Completely Macrostatistical Definition of Time Direction

Let us summarize the results to which our investigation concerning
time has thus far led us. We first defined time order in terms of mechanics. We then defined time direction in terms of thermodynamics, or

microstatistics. Finally, we applied the methods of microstatistics to macrostatistics and showed that time direction can be defined in terms of macrostatistics. This last definition merely reiterates relations which we had formulated as the hypothesis of the branch structure and which apply both to microstatistics and macrostatistics, including relations controlling the storage of information.

We shall now show that macrostatistics can even be used to define time order; that these statistics can replace the use of classical mechanics for the construction of a causal net which possesses a lineal order. The step from the lineal order of time to the direction of time is then easily made by the use of the principle of the common cause, as explained in §19. Before turning to the construction of the causal net, which may be illustrated by figure 5, §5, we must make some preliminary remarks concerning the observables presupposed in the construction.

We shall assume that events x, y, . . . are observed as separate spatiotemporal units. In order to construct a causal net for them by the use of statistical relations alone, we assume that the events x, y, . . . can be classified into certain classes A, B, . . . These classes, of course, are defined by us, but we impose upon their definition a certain condition.

We say that a class A is *codefined* if it is possible to classify an event x as belonging to A coincidently with the occurrence of x. In other words, observing x we must be able to say whether x belongs to A, and it must be unnecessary to know, for the purposes of this classification, whether certain other events y, z, . . . occurred earlier or later, or simultaneously at distant places. For instance, the class term "green" is codefined because looking at x we can say whether it is green. However, the class term "day following a rainy day" is not codefined; in order to know whether a day belongs in this class, we have to know whether it was preceded by a rainy day.

We now introduce for our statistics the requirement that the classes A, B, . . . be codefined. The use of codefined classes makes it possible to give reports of observations in terms of what is observed individually with respect to a spatiotemporal unit.

Logically speaking, a codefined class term is a one-place predicate which is not contracted from many-term predicates. For instance, a contracted term is given by the word "married", because "x is married" means as much as "there is a person y to whom x is married".[1] In conversational language, it is sometimes unclear whether a term is elementary or contracted. There is no such ambiguity in our present

---

[1] See Hans Reichenbach, *Elements of Symbolic Logic* (New York, Macmillan Co., 1947), p. 122.

investigation. The content of empirical observation is sufficiently distinct to enable us to say whether the statement "x belongs to A" is verifiable in terms of the observation of x alone.

We must now introduce a requirement concerning the probabilities referred to the defined classes. A probability expression has a unique numerical value only if its reference class is not empty.[2] In general, the classes to be used will be so constructed that they are not empty. But it will be permissible occasionally even to use empty reference classes on the following condition: the probability values which are used are derived from a set d of statements about observational data by means of a set r of rules, including inductive and deductive inferences. If we add to d the statement that the reference class is not empty, we require that the set d', thus resulting, be neither inconsistent nor contradict physical laws, and that if the rules r are applied to d', they lead to the same value for the probability. If this requirement is satisfied, we speak of *genuine* probabilities. It follows from the definition of this concept that it includes probabilities possessing nonempty reference classes.[3]

The construction of the causal net connecting the events $x, y, \ldots$ in terms of the classes $A, B, \ldots$ is achieved by means of a relation *causally between*. This relation is defined in statistical terms as a relation between classes. It can be applied to events as follows. We assume that triplets of events x, y, z occur repeatedly, and we wish to define a causal between-relation within each triplet. This includes the assumption that we can tell which events belong to the same triplet. The events of a triplet will not be too far apart, and we may be able to make a rough estimate of space-time distances. We do not want to compare directly events from different triplets. Such a comparison could only be constructed by means of intermediate events from other classes. Let us now imagine that we consider a certain triplet and find that x belongs to A_1, y to A_2, z to A_3. If we found out earlier that A_2 is causally between A_1 and A_3, we then say that y is causally between x and z. This interpretation is to be understood for the formulations to be given, in which we shall refer to classes only.

Assume that we have, in the usual language of positive time, a causal chain as represented in figure 30, in which positive time is indicated by arrows. If we know that x belongs to A_1, we then can predict with the probability $P(A_1, A_2)$ that y will belong to A_2. We abbreviate this by saying: if we know that A_1 has occurred, we can

---

[2] *ThP*, p. 55.

[3] Cf. Hans Reichenbach, *Nomological Statements and Admissible Operations* (Amsterdam, North-Holland Publ. Co., 1954), Appendix, p. 128.

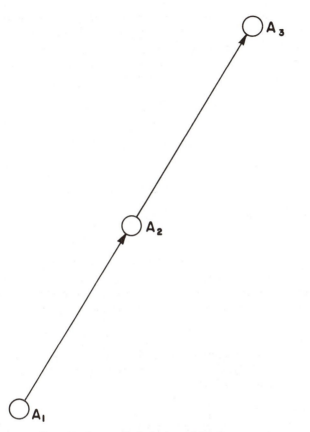

Fig. 30. A causal chain formed by three events.

predict with the probability $P(A_1, A_2)$ that A_2 will occur. Likewise we can predict that A_3 will occur. We will assume that the probability of the prediction is not equal to 1, but increases toward 1 as we come closer to the event considered; for instance, from A_2 we predict A_3 with a higher probability than that attainable from A_1. This is a relationship which holds quite generally for continuous probability sequences.[4]

However, once we know that A_2 has occurred, it is no longer necessary to know that it was preceded by A_1; event A_1 is no longer relevant to the prediction of A_3. The contribution of A_1 to A_3 has been absorbed in A_2, so to speak; and A_2 may be said to *screen off* A_1 from A_3. This will be true, at least, if the definitions of the classes are narrow enough

---

[4]*ThP*, p. 237.

to include a reference to observable traces. The assumption that definitions of this kind can be given may be regarded as formulating the principle of *action by contact.* We now define:

DEFINITION. An event A_2 is *causally between* the events A_1 and A_3 if the relations hold:

$$1 > P(A_2, A_3) > P(A_1, A_3) > P(A_3) > 0 \ , \tag{1}$$

$$1 > P(A_2, A_1) > P(A_3, A_1) > P(A_1) > 0 \ , \tag{2}$$

$$P(A_1 \cdot A_2, A_3) = P(A_2, A_3) \ . \tag{3}$$

We symbolize this relation in the form

$$\text{btw}(A_1, A_2, A_3) \ . \tag{4}$$

The expression (4) is a statement; it says that the events A_1, A_2, A_3 satisfy relations (1)–(3).

It can be shown that the relation *causally between* is *symmetrical* in A_1 and A_3, that is, that we can go from (4) to the statement

$$\text{btw } A_3, A_2, A_1 \ . \tag{5}$$

First, we see that relations (1) and (2) are merely exchanged when we interchange A_1 and A_3. Second, we use the general relation, proved in the theory of probability,[5]

$$\frac{P(A_i \cdot A_k, A_m)}{P(A_i \cdot A_m, A_k)} = \frac{P(A_i, A_m)}{P(A_i, A_k)} \ . \tag{6}$$

Putting here $i = 2$, $k = 1$, $m = 3$, we derive from (3) by the help of (6):

$$P(A_3 \cdot A_2, A_1) = P(A_2, A_1) \ . \tag{7}$$

This proves the symmetry. It follows that relations (1)–(3) do not supply the direction of the arrows in figure 30. The between-relation merely orders the three events.

The first inequality in (1) and the first inequality in (2) are required, because otherwise relations (7) and (3) would be trivially satisfied.[6] For instance, from $P(A_2, A_1) = 1$ we could immediately derive (3), because the addition of the factor "A_1" in the first term of the probability

---

[5] *Ibid.*, p. 112, equation (33).
[6] See *ibid.*, p. 117, equation (5).

expression "$P(A_2, A_3)$" does not change the value of $P(A_2, A_3)$ if $P(A_2, A_1) = 1$. It can be shown that also $P(A_1, A_2) < 1$ and $P(A_3, A_2) < 1$. If we had $P(A_1, A_2) = 1$, we could infer, using the same theorem, that $P(A_1, A_3) = P(A_1.A_2, A_3)$; when we then apply (3) we would find a contradiction to the second inequality in (1). A similar inference proves the second relation. Therefore these two relations need not be explicitly stated. We may add the remark that the third inequality in (2) follows from the third inequality in (1), because of the following relation, which is proved in the theory of probability:[7]

$$\frac{P(A_i, A_k)}{P(A_k, A_i)} = \frac{P(A_k)}{P(A_i)} \cdot \tag{8}$$

When we put $i=1$ and $k=3$, we conclude that, if the numerator on the left is larger than the one on the right, the same must hold for the denominators.

Furthermore, it can be proved that the relation *causally between* is *unique* for three events. This means that among three events only one can be causally between the two others. For the proof we have to show that, if (4) is true, the two relations

(9a) $\qquad$ btw $\left(A_3, A_1, A_2\right)$, $\qquad\qquad$ btw $\left(A_1, A_3, A_2\right)$, $\qquad$ (9b)

are false. The falsehood of (9a) follows from (6) when we put $i=1$, $k=2$, $m=3$, because the numerator on the left, according to (3) and (1), is then larger than the numerator on the right; therefore the denominator on the left must be larger than the denominator on the right, and we thus derive

$$P\left(A_3.A_1, A_2\right) > P\left(A_1, A_2\right) . \tag{10}$$

This shows that the analogue of (3) for the order A_3, A_1, A_2 is not satisfied. The falsehood of (9b) follows similarly from (6) when we put $i=3$, $k=2$, $m=1$ and use (7) and (2); we thus derive

$$P\left(A_1.A_3, A_2\right) > P\left(A_3, A_2\right) . \tag{11}$$

This relation contradicts the analogue of (3) for the order A_1, A_3, A_2.

Because of these properties, the relation *causally between* can be used to construct the causal net. The strict determination of the effect A_3 by its cause A_1, or A_2, which is formulated in the equations of mechanics and which we used in §5 to define a between-relation, is here replaced by a statistical determination. This statistical determination becomes weaker with growing causal distance, a fact which is

---

[7]*Ibid.*, p. 112, equation (32).

expressed in relations (1)–(2). And an intermediate event screens off
its predecessors from the effect, an occurrence formulated in relation
(3). The combined use of relations (1)–(3), therefore, leads to the order
relations holding for a causal net.

It should be noted that the applicability of the screening-off relation
(3) is dependent upon the use of codefined classes. Using other forms
of class definition, it would be easy to satisfy relation (3). For instance, when we define "A_2" as a class of events which practically
always are followed by A_3, we have $P(A_2, A_3) \sim 1$. The addition of
"A_1" in the reference term of the probability would then not make much
difference, and (3) would be easily satisfied. If statistical relations
are to inform us about causal order, they must refer to a classification
of events in which events are treated as individuals and their relations
to other events are not anticipated in the guise of class definitions.

When we have four events, the between-relations ascertained for
them permit us to distinguish between arrangements, such as are indicated in figure 31 and figure 32. For instance, if the following relations
have been verified,

$$\text{btw}(A_1, A_2, A_3), \quad \text{btw}(A_1, A_2, A_4), \quad \text{btw}(A_4, A_2, A_3), \quad (12)$$

we conclude that the four events have the order given in figure 32, and
not the order of figure 31.

We now must point out the limitations of the between-relation defined. The first limitation is that the between-relation formulated in
(1)–(3) is not transitive, that is, that we cannot prove the statement

$$\text{btw}(A_1, A_2, A_3) \cdot \text{btw}(A_2, A_3, A_4) \supset \text{btw}(A_1, A_2, A_4) \; . \quad (13)$$

The horseshoe sign means "implies". This statement may be true in
special cases, but it need not be true. The between-relation is therefore merely nontransitive, not intransitive. However, transitivity can
scarcely be required for a between-relation of this kind, which is used
to construct a net, and not a serial order. The occurrence of bifurca-
tions may lead to a violation of the inequalities (1)–(2). For instance,
in the arrangement presented in figure 23, §19, a prediction of E from
C may be more reliable than a prediction of E from B, although B is
geometrically between C and E along the line CBE. The reason is that
E is the common effect of B and A; and it may be possible to predict
A from C with a higher probability than attainable for A at B.

The relation *causally between*, therefore, merely pieces events together in a net, but does not establish a serial order, that is, an order
in one series. It may be called a *neighborhood relation;* but it is not a
serial between-relation.

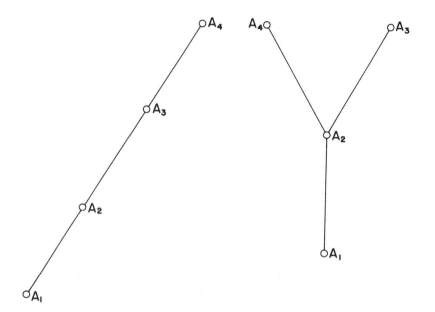

Fig. 31 (left) and fig. 32 (right). Two arrangements of events differing with respect to their causal between-relations.

The second limitation of the relation *causally between* concerns the relative direction of causal lines. Consider figures 33 and 34, in which positive time, as in figure 30, is indicated by arrows. Each of these arrangements can satisfy relations (1)–(3). The common cause, as well as the common effect, can screen off the event A_1 from A_3. For instance, if in figure 33 A_2 is the atmospheric pressure, A_1 the barometric reading, and A_3 rain, it is sufficient for the prediction of rain to know the atmospheric pressure; the knowledge that this pressure was recorded in A_1 is then redundant. However, the prediction of rain directly from A_2 is more reliable than the same prediction based on the record A_1 alone, because the barometer may be out of order. For the arrangement of figure 34, the illustration of the truck drivers from §19 may be used. The relation *causally between*, therefore, does not allow us to distinguish the arrangements of figure 33 and figure 34, in which the causal lines are *counterdirected*, from the arrangement of figure 30, in which the causal lines have the same direction, that is, are *equidirected*.

This result recalls a similar situation existing for the causal net of classical mechanics. We have seen in §5 that for two different mechanical motions, as illustrated in figure 3, the laws of mechanics could

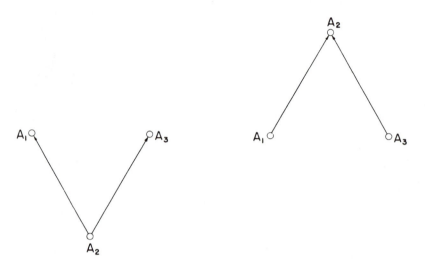

Fig. 33 (left) and fig. 34 (right). Both the common cause A_2 and the common
effect A_2 can be causally between two events A_1 and A_3.

not tell us their relative order. When we say that the two balls in
figure 3 either move in the direction of the solid arrows, or in the
direction of the broken arrows, this result is based on specific observations of coincidences and approximate coincidences, such as were
studied with respect to figure 4. We therefore introduced the principle
of the local comparability of time order, which formulates the methods
used for the discrimination of counterdirected from equidirected causal
lines. The same principle is required when this distinction is to be
carried out for a causal net based on statistical instead of strict
causal laws.

Consider an arrangement of the kind illustrated in figure 35. Assume
we find that A_2 is causally between A_3 and A_5, and that A_3 and A_5 are
in the neighborhood of A_2 and are still so close together that we can
call A_3A_5 an approximate coincidence. We then conclude that the
causal lines A_2A_3 and A_2A_5 are counterdirected. Likewise, assume that
A_2 is causally between A_1 and A_4, but that the combination A_1A_4 represents an approximate coincidence. We then infer that the causal lines
A_2A_1 and A_2A_4 are counterdirected. Furthermore, assume that A_2 is
causally between A_1 and A_3, but that A_1 and A_3 do not represent an
approximate coincidence. We then conclude that the causal line goes
in the direction $A_1A_2A_3$, or in the direction $A_3A_2A_1$, that is, that the
lines A_1A_2 and A_2A_3 have the same direction. Similar inferences can be
made for the line $A_4A_2A_5$.

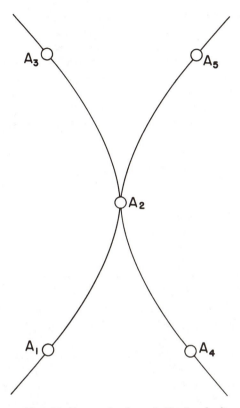

Fig. 35. Events in the neighborhood of
A_2, illustrating the local comparability of
time order.

These inferences are based on two presuppositions. The first concerns approximate coincidences. We must have observational means to find out that, for instance, A_3 and A_5 constitute a practical coincidence, whereas A_1 and A_3 do not form such a coincidence. Even if A_1 and A_3 are moved closer to A_2, we assume that, as long as these events can be distinguished from A_2, we can tell that they do not represent a practical coincidence. In other words, if we use a very inexact definition of "coincidence" such that A_1 and A_3 are combined into one event, then this event will also include A_2. It is different for the events A_3 and A_5, which for a certain inexact definition of "coincidence" constitute one event, whereas A_2 still forms a separate event.

The second presupposition concerns the inequalities (1) and (2). If the events A_1, A_2, A_3 are very close together, the first two probability expressions occurring in (1) and (2) are almost equal to 1, and it will be difficult to verify the difference in their values asserted in (1) and (2). This difference, however, is important, because it was needed for the proof of the asymmetry of the between-relation, expressed in (10) and (11). If we could not prove this asymmetry, the order $A_1A_2A_3$ could not be distinguished from the order $A_2A_1A_3$. For these reasons, we must presuppose that even for neighborhood events the inequality of the first two probability expressions in (1) and (2) can still be verified.

This presupposition, however, does not offer any difficulties. Although we may assume that, if A_1 occurs, A_3 is highly probable, it will be very different if A_2 does not occur; in this case, A_3 will be very improbable. This means that we can verify the relation

$$P\left(A_1 \cdot \bar{A}_2, A_3\right) < P\left(A_1, A_3\right) \ . \tag{14}$$

From this relation we can immediately infer[8] that

$$P\left(A_1 \cdot A_2, A_3\right) > P\left(A_1, A_3\right) \ . \tag{15}$$

Since the term on the left in (15) is equal to $P(A_2, A_3)$, according to (3), the inequality of the first two probabilities in (1) is now derivable. Similar considerations lead to the first two inequalities in (2). The asymmetry of the between-relation is thus preserved in the neighborhood of A_2.

When we wish to illustrate the application of the principle of local comparability of time order, we may use the example of the balls studied in connection with figure 4, §5. Since the relative probabilities occurring in (1) and (2) are almost equal to 1 if the events A_1, A_2, A_3 are close together, the statistical relations between the events are practically indistinguishable from causal laws and can be regarded as realized in mechanical processes. In fact, the use of strict laws for moving balls must be regarded as a schematization which is permissible if the probabilities are high, but which, in a more precise analysis, is to be replaced by probability relationships. The path of the moving balls is not strictly predictable, because possible disturbances by the environment cannot be completely ruled out. This problem was studied in §10; and it was shown there that actually attainable descriptions of mechanical processes can never supply absolutely certain predictions, but are limited to predictions of high degrees of probability. For these reasons, we may regard the inequalities (1)-(2) as satisfied by the moving balls, thus assuming that the relative probabilities connecting subsequent positions of the balls are never quite

---

[8]See *ThP*, p. 79, remarks following equation (11*b*). However, this inference is permissible only if $P(A_1, A_2) < 1$, which inequality was derived above from the relations (1)-(3).

equal to 1. The inequality (1) may then be regarded as proved by reference to the inequality (14), which latter relation can be illustrated by the example of the moving balls: if the ball is first at the place A_1, but is then prevented from arriving at the place A_2 by some disturbance from the outside, it will not arrive at the place A_3. In a similar way, the inequality (2) can be proved.

Once it is shown that the relations (1)–(3) are satisfied by the traveling balls, the discussion of observable coincidences can be taken over from the discussion given in §5.[9]

With the introduction of the principle of the local comparability of time order, the causal net constructed statistically has acquired the same properties as the causal net of classical mechanics: it is ordered as a whole. The net thus possesses a lineal order. This means that, if a time direction is assigned to one causal line, a direction results for every line. For the assignment of a direction, we can use the definition of time direction supplied by the principle of the common cause, given in §19. Thus the direction of time is constructed for the causal net as a whole. This definition of time direction is based on probability methods alone, and can be carried through completely within the frame of macrostatistics.

The given construction of the causal net and its direction rests on the assumption that there are events satisfying relations (1)–(3) and satisfying the principle of the local comparability of time order. In the following section, we shall study a statistical definition of time direction which is not subject to these conditions.

● 23. *The Mark Principle and Causal Relevance*

It is possible to construct the causal net and its direction by a direct use of irreversible processes, which are applied in such a way that they do not presuppose a previous order, but supply both an order and a direction. Because of its reference to irreversible processes, this construction makes use of microstatistics (see §17).

We shall assume, as in §22, that macroevents are given as spatio-temporal units and that they can be classified in codefined classes. The order relations between these events, however, will not be defined

---

[9]That A_2 screens off A_3 from A_5 has here the following meaning. If the ball arrives at A_3 we infer, first, that it was previously at A_2. Now we assume that it is known that at A_2, if this event took place at all, there was a collision with the other ball. Then we can conclude that the other ball will be at A_5. However, if the collision at A_2 was observed, it is not necessary, for the latter conclusion, to know that the first ball has arrived at A_3.

in terms of macrostatistics, but by the use of certain irreversible processes superimposed upon the existing statistical relations between
these events. This is achieved by means of a device called a *mark*.

A mark is the result of an intervention by means of an irreversible
process. We saw in §6 that a process may be differently interpreted,
according as it is regarded as having been initiated by an intervention
or by an alternative cause. If a reversible process is used for an intervention, therefore, we cannot use this process for the definition of
time direction; using the concept of alternative cause, we can interpret
the same process as going in opposite time direction. It is different
when we use an irreversible process for an intervention. Since the
reverse process does not occur, we cannot interpret the occurrence in
terms of an alternative cause, and thus we may use the intervention
for the definition of positive time.[1]

An intervention by means of irreversible processes can be deliber-
ately performed; it can also be the product of natural causes. Suppose
a white light ray goes from *A* to *B* (fig. 36); we perform an intervention
by putting a red glass in its path. When the light ray arrives at *B*, it is
then colored red, whereas there is no change at *A*. We say: the intervention by means of the red glass changes the light ray between the
glass and point *B*, but leaves the light unchanged between the glass
and point *A*.

Describing the same two processes in opposite time, we would say:
if no glass is there, white light travels from *B* to *A*; but the arrival of
white light in *A* can also be produced by an alternative cause: there is
red light traveling from *B* toward a red glass inserted in the path; this
glass emits light waves of different colors which are added to the
red beam in such a way that the resulting mixture is white and travels
toward *A*. A process of this kind, however, is never observed in positive time, because the absorption of light in a red glass is an irreversible

---

[1]In previous publications I have repeatedly used the mark principle for the distinction
of cause and effect: see Hans Reichenbach, "Bericht über eine Axiomatik der Einsteinschen Raum-Zeit-Lehre", *Phys. Zeitschr.*, Bd. 22 (1921), pp. 683-686; *Axiomatization of the Theory of Relativity* (Berkeley and Los Angeles, University of California
Press, 1969) ; and *Philosophie der Raum-Zeit-Lehre* (Berlin and Leipzig, W. de Gruyter
& Co., 1928). Some of my critics have interpreted my presentations as meaning that a
mark need not be constructed by the use of irreversible processes. This is a misunderstanding. In my book *Axiomatization of the Theory of Relativity* I said that possibly
the use of a mark presupposes the use of irreversible processes. However, in the early
twenties the problem of time direction had not yet been sufficiently analyzed, and I
believed that there exist, for a definition of time direction, other statistical methods
than those expressed in the second law of thermodynamics. This view I have now
corrected; see §19, n. 3.

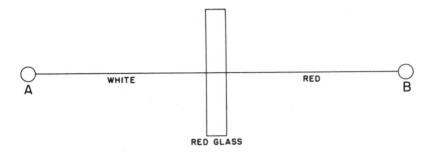

Fig. 36. Example of a marking process: insertion of a
red glass into a light beam.

process. Therefore we infer that the second description was given in
negative time. The intervention by means of the red glass thus defines
a direction of time, making the state at A the departure of the light
ray, or the cause, and the state at B the arrival of the light ray, or the
effect.

Other illustrations of marks are chalk marks on a billiard ball, the
smell of smoke in an air current, and snow on the roof of a railroad
car. All such processes are of the kind studied in the discussion of
the hypothesis of the branch structure. The act of intervention is the
branch-off point of the system and possesses a low entropy, or probability; the system then progresses to higher entropy, or probability,
and thus defines time direction. For instance, when the white light
ray hits the glass, this occurrence is an interaction, as considered by
Boltzmann, which starts the irreversible process of light absorption in
the glass. The glass gets warmer and progresses to higher entropy.
The red light ray is an ordered state constructed by selection from the
unordered mixture of rays of all colors; and thus it represents an
interaction product. When a piece of chalk touches the billiard ball,
grains of chalk are transferred to the ball; this is an irreversible process. Whenever we say, "The mark is being made at the event A", or
"The marking process occurs at A", we mean that the event A is a
point of interaction in which the lower end of a branch is situated. The
term "starting point of a marking process" refers to this point of interaction. The time direction ascertained by the use of a marking process
is therefore the direction of increasing entropy, expressed in the
branch structure of the universe.

If a marking process applied to an event A_i shows in another event
A_k, we know that A_i precedes A_k. Furthermore, we know that A_i is a
cause contributing to the existence of A_k; this relationship may be

called *causal relevance*. We therefore introduce the following definition:

DEFINITION 1. If a mark made in an event A_i shows in an event A_k, then A_i is *causally relevant* to A_k.

When we wish to establish an order among the events $A_1 \ldots A_n$, it is not sufficient to make a mark in A_1. This mark may show in all the events $A_2 \ldots A_n$; but we do not thus know the order among these events, and we cannot even tell whether the events are arranged in one causal line. In order to find out causal lines we have to use marking processes repeatedly, starting first at A_1, then at A_2, and so on. Assume that a mark made at A_1 shows in A_2, and perhaps also in A_3, and assume that a mark made in A_2 shows in A_3; then we know that A_2 is causally between A_1 and A_3. Going on in the same way, we can order the events $A_1 \ldots A_n$ in a causal line, which has a direction, stemming from the irreversibility of the marking process.

That this procedure cannot lead back to the event A_1, but supplies open lines, follows from assumption 5 of the hypothesis of the branch structure, formulated in §16. If a line constructed by the marking process were closed, there would be at least one irreversible process which did not have the same direction as the others; and this is so improbable, in the sense of a many-system probability, that it may be disregarded.

Let us assume that we thus find that the events $A_1 \ldots A_n$ are arranged in a causal line. We then call the events $A_1 \ldots A_{n-1}$ the *causal ancestry* of A_n; and the events $A_2 \ldots A_n$ are called the *causal descendants* of A_1. The corresponding relations in a family tree are precisely of this kind. Marking processes are like tracers drawing causal lines between events. Incidentally, the radioactive tracers used nowadays to reveal the causal structure of circulation in living organisms represent processes indicating the existence of causal lines.

The marking process is sufficient to make possible the construction of the causal net, which is thus associated from the very beginning with a direction. However, it is also possible to combine the marking process with other statistical methods. For instance, we can first construct a causal net by the use of the relation *causally between*, defined in (22, 1-3). Instead of using the principle of the local comparability of time order, we can then use marking processes to discriminate between counterdirected and equidirected lines, and thus to find out causal lines of a continuous direction. The between-relations along these lines need not be tested by the use of further marks, but can be taken from the previous construction.

A mixed procedure of this kind, however, presupposes a certain correspondence between the order established by marking processes and the order derived from the relations (22, 1–3). This correspondence must now be studied. It is based on certain assumptions, which we shall formulate separately.

With reference to the last part of the inequality (22, 1) we introduce the following assumption:

Assumption α. If a mark made in A_i shows in A_k, then

$$P(A_i, A_k) > P(A_k) \ . \tag{1}$$

This inequality is thus assumed to be a necessary condition for the transfer of the mark, though it is not a sufficient condition. This assumption may be regarded as justified by many experiences. The inequality (1) formulates statistical relevance in the widest sense; and it appears plausible that the narrower concept of causal relevance presupposes the wider concept of statistical relevance. More precisely speaking, (1) expresses *positive relevance* in the statistical sense. If the larger-than sign were replaced by a smaller-than sign, we would have *negative relevance*. Only if it were replaced by an equality sign would we have *irrelevance*. We may therefore interpret assumption α as meaning that causal relevance is a special form of positive relevance.

We must now introduce a second assumption concerning the marking process, denoting by A' the event resulting when the mark is added to A:

Assumption β. If a mark is made in A_i, then

$$either \ P(A_i', A_k') = P(A_i, A_k) \tag{2}$$

$$or \ \ P(A_i', A_k) = P(A_i, A_k) \ . \tag{3}$$

If (2) holds we say that the mark made at A_i shows in A_k. If (3) holds we say that it does not show in A_k. The latter case will always occur if A_k precedes A_i; but it can also occur if A_k follows A_i, because the mark may vanish at some place along the causal line leading to A_k. For instance, the snow on the top of a railroad car may melt. Equations (2)–(3) state that the marking process is controlled by relations of practical certainty; it either occurs, or does not occur, at the second event, depending on the kind of mark used, and thus does not change the existing probabilities.

We now must investigate the connection between the marking process and the screening-off relation defined in (22, 3). When the marking process is used to indicate causal lines, the mark shows in all events along the line; if the mark is stopped at some place, it does not reappear at later places. For instance, at a certain place a chalk mark on a billiard ball may be erased; if so, it does not show on the ball from there on. This consideration leads to the following assumption, which expresses the *continuity of the mark:*

Assumption γ. If A_2 screens off A_1 from A_3, and if a mark made in A_1 shows in A_3, then it also shows in A_2.

This assumption must be regarded as an empirical hypothesis, according to which the marking process conforms to the principle of action by contact. We can adduce the following argument in its favor. We can prove that, unless assumption γ is true, the screening-off relation is disturbed by the marking process, that is, does not hold any longer when the marking process is carried out. In other words, assumption γ is a necessary condition if we assume that the marking process does not disturb the screening-off relation.

In order to give this proof, let us assume that the mark made in A_1 shows in A_3, but not in A_2. We want to show that this assumption is incompatible with the assumption that the screening-off relation holds for these events both with and without mark. The latter assumption is expressed by the two relations:

$$P(A_1 . A_2, A_3) = P(A_2, A_3) \ , \tag{4}$$

$$P(A_1' . A_2, A_3') = P(A_2, A_3') \ . \tag{5}$$

We will prove that these two relations contradict the relations (2)–(3).

We begin by using the rule of elimination,[2] a theorem of the calculus of probability, in the form

$$P(A_1', A_3') = P(A_1', A_2) \cdot P(A_1' . A_2, A_3') + P(A_1', \bar{A}_2) \cdot P(A_1' . \bar{A}_2, A_3') \ . \tag{6}$$

Applying (2) to the term $P(A_1', A_3')$, and (3) to the term $P(A_1', A_2)$, we find, using (5):

$$P(A_1, A_3) = P(A_1, A_2) \cdot P(A_2, A_3') + \left[1 - P(A_1, A_2)\right] \cdot P(A_1' . \bar{A}_2, A_3') \ . \tag{7}$$

Since A_3' can occur only if A_3 does not occur, that is, if $\bar{A}_3$ occurs, we have

$$P(A_2, A_3') \leq 1 - P(A_2, A_3) \ . \tag{8}$$

The value of the last probability expression in (7) is unknown; when we replace it by its highest possible value, which is equal to 1, we can only make the right-hand side of (7) larger. We thus find, using (8):

---

[2]*ThP*, p. 76.

$$P(A_1, A_3) \leqq P(A_1, A_2) \cdot \left[1 - P(A_2, A_3)\right] + 1 - P(A_1, A_2) ,$$

$$(9)$$

$$P(A_1, A_3) \leqq 1 - P(A_1, A_2) \cdot P(A_2, A_3) .$$

When we now apply the rule of elimination to the events A_1, A_2, A_3, we arrive at an equation analogous to (6), the only difference being that the prime marks are omitted. Using (4) we derive:

$$P(A_1, A_3) \leqq P(A_1, A_2) \cdot P(A_2, A_3) . \tag{10}$$

Adding the two inequalities (9) and (10), we find that they lead to the inequality

$$P(A_1, A_3) \leqq \tfrac{1}{2} . \tag{11}$$

Now we can derive from (4)–(5) the two relations:

$$P(A_1 . A_2, \overline{A}_3) = P(A_2, \overline{A}_3) , \tag{12}$$

$$P(A'_1 . A_2, \overline{A}'_3) = P(A_2, \overline{A}'_3) . \tag{13}$$

Furthermore, we derive from (2)–(3) the two relations,

$$\text{either } P(A'_i, \overline{A}'_k) = P(A_i, \overline{A}_k) \tag{14}$$

$$\text{or } \quad P(A'_i, \overline{A}_k) = P(A_i, \overline{A}_k) . \tag{15}$$

When we now construct from relations (12)–(15) a proof exactly like the preceding one, we find

$$P(A_1, \overline{A}_3) \leqq \tfrac{1}{2} . \tag{16}$$

This inequality contradicts (11), except when $P(A_1, A_3) = 1/2$. However, we can exclude this case by the following consideration. The equality sign in (9) is possible only if $P(A'_1 . \overline{A}_2, A'_3) = 1$. But since, in that case, $P(A'_1 . \overline{A}_2, \overline{A}'_3) = 0$, we can drop the equality sign in (16) under these conditions. It is easily seen that because of this relationship only one of the relations (11) and (16) can require the equality sign. But then these inequalities are always mutually contradictory. This concludes the proof that, if the marking process does not disturb the screening-off relation, assumption γ must hold.

We shall now consider a certain extension of the screening-off relation. In an arrangement as drawn in figure 35, it may happen that neither $A_2^{(1)}$ nor $A_2^{(2)}$ screens off A_1 from A_3, whereas their conjunction

$A_2^{(1)}.A_2^{(2)}$ does so. This means that if we put "$A_2^{(1)}.A_2^{(2)}$" for "A_2," in relation (22, 3), this relation is satisfied. We can therefore speak of a set $A_2^{(1)} \ldots A_2^{(n)}$ screening off A_1 from A_3. Assumption γ may now be extended as follows:

Assumption δ. If a set $A_2^{(1)} \ldots A_2^{(n)}$ screens off A_1 from A_3, and if a mark made in A_1 shows in A_3, then it also shows in at least one of the events $A_2^{(1)} \ldots A_2^{(n)}$.

The assumptions α–δ formulate the correspondence between the marking process and the relation *causally between*, defined in (22, 1–3). Using this correspondence, we can now introduce a second definition of causal relevance, which is equivalent to the first and is applicable when time direction has already been defined. Since this presupposition is satisfied for many practical applications, the second definition of causal relevance supplies a convenient tool of great practical use.

Assume that none of the events $A_2^{(1)} \ldots A_2^{(n)}$ is later than A_1; then some of these events may be earlier than A_1, others may be simultaneous with A_1. Since we know that a mark always travels forward in time, a mark made in A_1 cannot show in any of the events $A_2^{(1)} \ldots A_2^{(n)}$. Let A_3 be an event which is screened off from A_1 by the set $A_2^{(1)} \ldots A_2^{(n)}$; we then derive from assumption δ that a mark made in A_1 cannot show in A_3. Combining this result with assumption α, we arrive at the following definition:

DEFINITION 2. An event A_1 is *causally relevant* to a later event A_3 if

$$P(A_1, A_3) > P(A_3) \tag{17}$$

and there exists no set of events $A_2^{(1)} \ldots A_2^{(n)}$ which are earlier than or simultaneous with A_1 such that this set screens off A_1 from A_3.

This definition of causal relevance, which presupposes time direction, satisfies the following condition: If A_1 is causally relevant to A_3, then the transfer of a mark from A_1 to A_3 is not excluded by the continuity assumption δ. Moreover, that in all such cases of causal relevance it is physically possible to find or construct a mark proceeding from A_1 to A_3 is an assumption which may be regarded as confirmed by many experiences. The extensional identity of the two definitions of causal relevance is therefore a matter of empirical evidence.

The second definition of causal relevance can be used where a marking process cannot be employed for technical reasons, as, for instance, in cases exempt from experimental intervention. We can briefly state the leading idea of the definition in the form: If an event A_1 is screened off from a later event A_3 by an event A_2 which is not later than A_1, then A_1 is not causally relevant to A_3. However, if A_2 were later than A_1, it might screen off A_1 and yet leave A_1 causally relevant to A_3.

Both definition 1 and definition 2 of causal relevance are so constructed that they express a direction and thus formulate *causal influence* as distinguished from *causal indication*. If A_1 is causally relevant to A_3, then A_1 contributes to the existence of A_3. The vague terms "causal influence", and "contributing to the existence of an event", are given a precise explication by reference to the transfer of a mark, or to the absence of an event A_2 earlier than A_1 which screens off A_1 from A_3. In contrast, causal indication can be identified with positive statistical relevance in general, as formulated in (1).

CHAPTER V THE TIME OF QUANTUM PHYSICS

• 24. The Statistical Reversibility of the Elementary Processes of Quantum Mechanics

For the classical statistics of atoms, the problem of the direction of time has been shown to present itself in a paradoxical form: the irreversibility of macroprocesses is to be reconciled with the reversibility of microprocesses.[1] This paradox of the statistical direction was solved, in a continuation of Boltzmann's ideas, by the recognition of the sectional nature of time direction: a large isolated system can indeed define a time direction in a section of its whole temporal development, if this section is rich in branch systems governed by the laws of statistical isotropy.

Boltzmann's problem sprang from the conception that atomic processes were controlled by the laws of Newton's mechanics, which do

---

[1]See pp. 109, 117, and 134.

not distinguish one time direction from the opposite one, a conception which implies the reversibility of all elementary processes. When the discoveries of quantum physics led to the conclusion that atomic processes cannot be incorporated into classical mechanics, that they require the construction of a specific quantum mechanics which in all its essential features is very different from Newton's conceptions, many physicists were hopeful that the new mechanics might allow a different solution of the problem of time direction. The possibility was envisaged that the elementary processes of quantum mechanics were not subject to temporal symmetry. If this conjecture had been true, the direction of time would be immanent in the most fundamental laws of physics, and· we would not have to resort to statistical considerations in order to account for a temporal direction.

The hope for a solution of this kind was soon abandoned by the physicists. The study of the relations of quantum mechanics made it convincingly clear that quantum processes do not distinguish one time direction from its opposite, and that such processes are as reversible as those of classical mechanics. This is true even though quantum processes are governed not by strict causal laws, but by probability laws. It is a consequence of these results that the definition of time direction depends, for quantum physics just as much as for classical physics, on statistical properties of macrocosmic processes. Yet it will be seen that the combination of quantum laws for the microcosm with classical statistical laws for the macrocosm leads to a new aspect of time, because quantum physics is constructed on the basis of a causal indeterminism. In fact, the unidirectional nature of time assumes an even more pronounced form when it is conceived as the joint product of classical statistical properties of macroprocesses and an indeterminacy of microprocesses. The development of this thesis is the program of the present part of this book.

It must first be explained why quantum processes are reversible. In classical mechanics, the reversibility of the motion of mass points results from the fact that the differential equations of motion are of the second order with first derivatives absent; as a consequence, if $f(t)$ is a solution, $f(-t)$ is also a solution (see §4). In contrast, Schrödinger's time-dependent equation, which governs the time development of the ψ-function, is of first order with respect to the variable t. It might appear, therefore, as though this equation does distinguish one time direction from its opposite.

To make this consideration clearer, let us remark that in quantum mechanics a physical quantity u is characterized by its probability distribution, that is, by a function $d(u)$ which tells us with what probability a certain value of the quantity will be observed, if a suitable measurement is made. The function $\psi(q)$, which occupies a central place in the mathematics of quantum mechanics, can be regarded as a mathematical device to collect all these distribution functions, for all quantities involved, in one function. The quantity q stands for the set of position parameters of the system considered; for a mass point, q is simply the set of its three spatial coördinates. The values ψ of this function are complex, although its arguments q are real numbers. The various distribution functions $d(u)$, which are real functions, being probability densities, are derivable from $\psi(q)$ by certain mathematical rules which need not be studied here.[2]

The function $\psi(q)$ thus supplies a description D of the physical state of the system; this description is as complete as is at all possible. When the state changes in the course of time, the variable "t" enters as another argument into the function, which is then written in the form "$\psi(q, t)$". For the time development of this function, Schrödinger has constructed a differential equation which expresses the fundamental law of change in quantum mechanics. This equation has the form

$$H_{op}\psi\big(q,\, t\big) = c\, \frac{\partial \psi(q,\, t)}{\partial t} \,, \qquad c = \frac{ih}{2\pi} \,. \tag{1}$$

The symbol "H_{op}" expresses a mathematical operator derived from the energy H. An operator is a rule by which to construct a certain derivative function from the function to which the operator applies. The direction of time, that is, the temporal direction in which the change occurs, manifests itself in the sign of the argument "t".

To study this problem, we have to go into some mathematical details which will be understandable only to a reader who is familiar with the mathematical technique of quantum mechanics. However, one can still understand the conclusion without being acquainted with this technique.

We remark first that, if $\psi(q, t)$ is a solution of this equation, $\psi(q,-t)$ is not a solution. The latter function satisfies instead the equation

$$H_{op}\psi\big(q,\, -t\big) = -c\, \frac{\partial \psi(q,\, -t)}{\partial t} \,, \tag{2}$$

which differs from (1) in the minus sign on the right-hand side.

There remains the problem of distinguishing between $\psi(q, t)$ and $\psi(q, -t)$. In order to discriminate between these two functions, we would first have to know whether (1) or (2) is the correct equation. But

---

[2]For an introduction to quantum mechanics which is suitable to explain in more detail the considerations mentioned in the present chapter, see *PhF* [*Philosophical Foundations of Quantum Mechanics*].

the sign of the term on the right in Schrödinger's equation can be tested observationally only if a direction of time has been previously defined. We use here the time direction of the macrocosmic systems by the help of which we compare the mathematical consequences of Schrödinger's equation with observations. Therefore, to attempt a definition of time direction through Schrödinger's equation would be reasoning in a circle; this equation merely presents us with the time direction which we introduced previously in terms of macrocosmic processes.

It may be recalled that even in classical physics the time direction of a molecule is not ascertainable from observations of the molecule, even if such direct observations could be made, but is determined only by comparison with macroprocesses, for which statistics define a time direction. In the same way, the time direction of a quantum-mechanical elementary process, like the movement of an electron, is determined only with reference to the time of macroprocesses.

This consideration shows that the fundamental quantum-mechanical law governing the time development of physical systems does not distinguish one time direction from its opposite. Since the laws governing the observables of quantum physics are not causal laws, but probability laws, the reversibility of elementary processes assumes here the form of a symmetry of the relations connecting probability distributions. These connecting relations are strict laws expressible through a differential equation, namely, Schrödinger's equation (1); the statistical laws governing observables are thus tied together by a strict mathematical linkage which controls the time change of the probability distributions.

The existence of this linkage has sometimes been construed as a return to causal laws. But there is a difference between "causal law" in its traditional sense and in the sense of a mathematical relation between probability distributions. Conceived as relations between observables, there are only statistical laws in quantum mechanics. When we say that the ψ-function describes a physical state, we must not forget that what we call here a state description represents merely a statistical description of observables. It tells us with what probability our measurements will furnish any particular results. And a strict linkage between probability distributions does not make the relation between observables a strict relation. In other words, if we go from observables A to a distribution B in terms of a probability implication, from B to another distribution C in terms of a strict implication, and from C to observables D in terms of a probability implication, then the relation between A and D is still a probability implication.

But it is true that the time change of physical states is expressible as a strict law governing the linkage. Using this conception, we can describe the temporal symmetry of quantum mechanics as follows. The function $\psi(q, t_1)$, extending over a certain area in the q-space, determines the set s_1 of probability distributions of all physical quantities involved in the system at time t_1; likewise, the function $\psi(q, t_2)$ determines the set s_2 of these probability distributions at the time t_2. If a physical law admits the transition from s_1 to s_2, then the inverse transition from s_2 to s_1 is also admitted by a physical law. In this version, the reversibility of elementary processes reappears in quantum physics.

It is not possible to assert a direct reversibility of every elementary process, since not all such processes are governed by strict laws. If we cannot say that a process, on repetition of the same initial conditions so far as they can be specified, will arrive at the same result for the same direction of time, we cannot say, either, that the reverse process would lead back to exactly the same physical conditions. Instead, we can only speak of a *statistical reversibility*. This statistical reversibility is expressed in the form of a temporal symmetry of the strict law determining the time development of probability distributions. Thus the problem of time direction is left by quantum mechanics in the same status as that in which it presented itself to classical physics.

We must now examine the changes which quantum mechanics has introduced for the problem of determinism.

• *25. The Indeterminism of Quantum Mechanics*

The indeterminism of quantum mechanics has been expressed in W. Heisenberg's relation of indeterminacy. The conceptual system of quantum mechanics led to the conclusion that for every physical quantity u there exist many other quantities v which cannot be measured simultaneously with u. Since this relationship is expressed, in the algorism of quantum mechanics, by the noncommutativity of the product of the matrices belonging to these quantities, such quantities are called *noncommutative*, or *noncommuting*.[1] This does not mean simply a division of all quantities into two classes; the division separates many more classes, because there are quantities u, v, w, none of which commutes with any of the others. A particular instance of noncommuting quantities is given by position q and momentum p of a particle.[2]

---

[1] By contrast, the usual arithmetic product is always commutative: $x \cdot y = y \cdot x$.

[2] It is common usage to denote the momentum by the letter "p" and the position by the letter "q". Although I use the same letters to denote the values of probabilities or similar quantities, the intended meaning will always be clear from the context.

Heisenberg has shown that the values q and p can be measured only to a certain exactness Δq and Δp so as to satisfy the inequality:

$$\Delta q \cdot \Delta p \geqq \frac{h}{4\pi} \; . \tag{1}$$

The quantity h, Planck's quantum of action, is very small; therefore the inexactness shows only for very precise measurements. And since $p = m \cdot v$ for a mass point, the corresponding relation for position q and velocity v assumes the form:

$$\Delta q \cdot \Delta v \geqq \frac{h}{4\pi m} \; . \tag{2}$$

If m is large, the right-hand side is very small. For particles of large mass, such as macroscopic particles, or dust particles visible under the microscope, there thus is practically no limitation of measurability.

For particles like electrons and the many other microparticles discovered in quantum physics, however, the restriction (1) is noticeable. This restriction, still, leaves us a choice; we can measure either q or p as exactly as we wish, but the other quantity then is measured very inexactly. A similar restriction applies to any pair of noncommutative quantities.

This limitation of measurability leads to a limitation of predictions. It was shown in §11 that a precise prediction of the future state of a physical system requires a precise knowledge of its present state. If this knowledge is subject to the limitation expressed in (1), a precise prediction is not possible. It should be noticed that an exact measurement of one quantity is not even sufficient to permit prediction of later values of that same quantity. For instance, even though we measure the position of a particle very precisely, we cannot predict its future position unless we also know the value of its velocity, or momentum, at the present moment.

Using the terminology developed in §11, we may say that Heisenberg's relation does not directly exclude predictions in a conditional form, but does exclude predictions in an unconditional form. And predictions of the latter kind are the subject of determinism. It is not possible, according to Heisenberg, to construct for the present physical state a description $D^{(i)}$ as precise as is needed if we wish to predict a future state with a probability $p^{(i)}$ close to 1. Consequently, both schema (11, 1) and schema (11, 2) must be replaced, for quantum mechanics, by the schema

$$D^{(1)} \quad D^{(2)} \quad D^{(3)} \quad \ldots \ldots D$$

$$\overset{(1)}{p} \quad \overset{(2)}{p} \quad \overset{(3)}{p} \quad \ldots \ldots p \tag{3}$$

There thus exists an ultimate description D, the best we can construct; it is given by the ψ-function of the present physical state. But this description allows us to predict the future state only with a probability p lower than 1, and no means exist ever to improve this probability. Schema (3) expresses the indeterminism of quantum mechanics and the existence of a limit of predictability.

This formulation poses some important questions. Is the limitation for the human observer? Does there exist an exact value of the measured quantity, though unknown to us? And does there exist a precise description D', though unknown to us, which would permit a prediction with a probability p' much closer to 1 than p is, or even a probability p' equal to 1? This question can also be phrased in the following way: if we have to abandon the unconditional form of causality, must we also abandon its conditional form? Or can we admit, in principle, the existence of causal laws of the form: if the particle has now exactly the position q_1 and the momentum p_1, then, at a later time, its position will be q_2, and its momentum p_2?

This question is sometimes raised in an even more optimistic version. It is argued that the limitation of measurability may apply merely to the knowledge of our time, and that at a later time methods may be found by which this limitation may be overcome, and measurements of noncommuting quantities may be made as exactly as desired. At this later time, the schema (3) would no longer express the predictive situation. Thus this schema would only characterize the physical science of our time, but could not be regarded as expressing a property of the physical world.

The answer to these questions is not easily given. It requires a logical analysis of the meaning of statements about the physical world, of the nature of the inferences that lead from statements about observables to statements about unobservables. And it presupposes a study of the considerations that have led to Heisenberg's inequality, an inquiry that can help us to stake out the range of this relation, either to the effect that it expresses a physical law or merely that it constitutes a limitation of the capacities of human observers at the present time. It will be seen that the latter alternative is not compatible with an unprejudiced analysis of quantum-mechanical results.

Let us examine first how the indeterminacy relation (1) is derived within the system of quantum mechanics. We recall that the function $\psi(q)$ represents a mathematical device by which to unite, in one function, all probability distributions $d(u)$ referring to physical quantities involved in the system (see §24). The individual distributions $d(u)$ are derived from $\psi(q)$ by certain rules laid down in the mathematics of quantum mechanics. In particular, we can derive from $\psi(q)$ the two distributions $d(q)$ and $d(p)$ for position and momentum. We use here the symbol "d" as standing for "distribution"; the mathematical functions $d(q)$ and $d(p)$ can be of very different forms. Now it can be shown that, because of their origin from one function $\psi(q)$, the two functions $d(q)$ and $d(p)$ are connected by an inverse correlation: if $d(q)$ is a steep curve, then $d(p)$ is a flat curve, and vice versa (see fig. 37).[3] This inverse correlation is derivable whatever be the specific form of the function $\psi(q)$. When we understand by Δq and Δp the standard deviations belonging to the curves, we thus arrive at Heisenberg's indeterminacy relation (1).

This relation is therefore deeply anchored in the system of quantum mechanics. Yet it is not founded any better than this system itself. It is contingent, in particular, upon one general assumption on which all conclusions of quantum physics are based: the assumption that all observational knowledge ever attainable can be summarized in the form of a ψ-function; in other words, that the description D of a physical state given by the ψ-function is the most complete description possible. We shall call this assumption the *synoptic principle* of quantum mechanics. If this principle is true, Heisenberg's indeterminacy is inescapable, because it is a mathematical consequence of the principle. But if the principle should not be true, the indeterminacy relation may be abandoned.

The reason is that the indeterminacy principle includes a negative assertion: it states that there are no observables that would lead to better measurements than those compatible with the indeterminacy relation (1). The mathematical derivation of this relation refers only to quantities governed by the ψ-function; any two noncommuting quantities governed by the ψ-function satisfy the inequality (1). If it is claimed that this inequality holds quite generally, then it is assumed that there are no observables that could not be included in a ψ-function. With its extension to all observables, Heisenberg's relation of indeterminacy presupposes the synoptic principle.

It is important to keep this logical relationship in mind. We can speak of a proof of Heisenberg's relation only when we regard the

───────────

[3] *PhF*, p. 11.

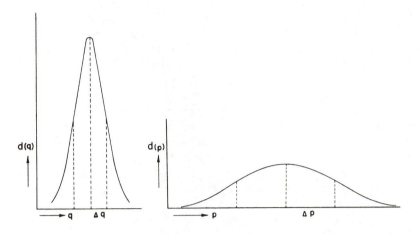

Fig. 37. Inverse correlation of position distribution $d(q)$ and momentum distribution $d(p)$, which are both derived from the function $\psi(q)$. Distribution of the position q is in the form of a Gauss curve.

Compare with *PhF*, page 9, figures 2 and 3, from which this illustration is adapted. Curve $d(p)$, above, is constructed through the Fourier expansion of a ψ-function from which the curve $d(q)$ is derivable; for the direct Fourier expansion of curve $d(q)$, see the dotted-line curve in figure 3 of *PhF*.

synoptic principle as valid; whatever proof has been presented in quantum theory for the relation of indeterminacy hinges upon the acceptance of the synoptic principle. This qualification applies, in particular, to a proof given by Von Neumann,[4] which reiterates Heisenberg's result in a mathematically generalized version, but is conclusive only if the synoptic principle is true. An inquiry into the validity of quantum-mechanical indeterminism, therefore, must concern itself with the evidence available for the synoptic principle.

There are two kinds of evidence for this principle. One kind is negative: no physical system has ever been found in which the synoptic principle is not satisfied. If this were all the evidence we had, the principle would be poorly founded. Fortunately, there also exists evidence of a positive kind: the whole system of quantum mechanics calls for the validity of the principle, because without the principle this system would lead to absurd consequences.

Before entering into a discussion of these consequences, let us illustrate the logical structure of the argument by an example taken

[4]John von Neumann, *Mathematische Grundlagen der Quantenmechanik* (Berlin, J. Springer, 1932), p. 170.

from the theory of relativity. Einstein's principle of the limiting character of the velocity of light is based both on negative and on positive evidence. It is negative evidence for the principle that no signals faster than light have ever been observed. But Einstein's principle would never have been accepted, had it not been based, in addition, on positive evidence. The special theory of relativity would lead to absurd consequences if the principle were not true; the time order of causal processes could be reversed, infinite amounts of energy could be produced, and so on. Consequences of this kind appear so unlikely that they supply an argument in reverse: it appears very unlikely that the principle of the limiting character of light velocity is false. The improbability of consequences that would invalidate the principle is thus turned into inductive evidence for it.

It is this form of inference which we can use to construct positive evidence for the synoptic principle of quantum mechanics. The evidence thus adduced is, of course, of an inductive nature; in other words, there exists no absolute proof for the principle. It can only be supported by reference to general experience and represents an empirical hypothesis. But the positive experiential support is very strong, because it is based on very general features of the physical world.

In order to understand the positive evidence existing for the synoptic principle we must consider another fundamental principle of quantum mechanics, which, like Heisenberg's principle, is derivable from the foundations of quantum mechanics, but which does not presuppose the synoptic principle. For this other principle I have proposed the name of *principle of anomaly*. It is related to Bohr's principle of complementarity, but it goes beyond Bohr's principle inasmuch as it makes precise statements about the unobservables of quantum mechanics.

The observables of physics—and this includes quantum mechanics—are always macroscopic things. They are figures on dials, stripes on photographic films, and so forth. All statements about physical quantities of a smaller scale are derived by way of inferences based on these macroscopic observables.

These inferences proceed by steps. We first infer that some elementary coincidence has taken place, such as a collision between two electrons, or between a γ-ray and an electron. These inferences are based on a rather simple theory; in fact, this theory does not include quantum mechanics, but can be fully given in terms of classical conceptions. For elementary coincidences of this kind I have proposed the name of *phenomena*. They are sometimes called "observables" in a somewhat wider than usual sense of the term; to avoid confusion, we might call them "micro-observables", as distinguished from macro-observables. They function as observables do, in that they supply a basis for inferences toward another kind of occurrences.

These other occurrences are happenings that take place between the phenomena, for which happenings I have therefore used the name of *interphenomena*. The inferences leading to them can be made only within the framework of quantum mechanics. Furthermore, these inferences presuppose certain definitions, or *extension rules* of language, which allow us to extend the language of phenomena to that of interphenomena and thus to speak meaningfully about occurrences which cannot be observed, even when "observing" is meant in the wider sense. Being definitions, these rules are not true or false; however, we can ask what consequences arise from the use of a language constructed by these rules.

The terms "particle" and "wave" both belong to the language of interphenomena. They assert something about what happens between the localized phenomena. Speaking of particles, we say that the interphenomena, too, are localized. Speaking of waves, we say that the interphenomena are spread over a wide area, although their final products, the phenomena, are localized. It can be shown that the extension rules leading to these two kinds of entities are rather simple. The result of a measurement is usually interpreted as the value of the measured quantity *after* the measurement. When we lay down the rule that the quantity had the same value *before* the measurement, we have introduced the particle interpretation. When we decide for the rule that the quantity before the measurement had all its possible values simultaneously, we have introduced the wave interpretation. Both interpretations are *exhaustive*, that is, they assign values, not only to phenomena, but also to interphenomena.

We can now ask for the physical properties of the interphenomena introduced by our definition. Here we are given the strange answer that the behavior of the interphenomena violates the principle of causality. By this violation I mean a deviation from the principle of action by contact. The further path of a particle that goes through a slit depends on whether some other slit is open or closed. A wave vanishes along its whole front when it produces a localized coincidence, a phenomenon, at one point of its front. And there are many other anomalies, some of which will be discussed in the next section. By the principle of anomaly I understand the statement that whenever we use an exhaustive interpretation, there result deviations from the principle of action by contact for the interphenomena. More precisely speaking, this principle can be formulated as follows:[5]

Principle of anomaly. If we assign, by any kind of rules, simultaneous values to noncommuting quantities the distributions of which are governed by the ψ-function, then the resulting interphenomena are subject to causal anomalies.

In view of this result, one may decide not to speak of interphenomena, but to restrict the language of quantum mechanics to phenomena. Such a *restrictive interpretation* can be carried out. Of this kind is the interpretation developed by N. Bohr and W. Heisenberg, who regard statements about interphenomena as meaningless. An interpretation in terms of a three-valued logic, which I have proposed, appears preferable, because it still permits the inclusion of interphenomena in some sense in the language of physics, although it is restrictive inasmuch as it excludes assertions about a definite status of the interphenomena. But these considerations are irrelevant to our present investigation. Being concerned with the problem of indeterminism versus determinism, we are interested in exhaustive interpretations.

Returning, therefore, to exhaustive interpretations, let us remark that thus far we have not made use of the synoptic principle. The principle of anomaly was formulated for quantities the distributions of which are governed by the ψ-function; but it did not include a presupposition that there are no quantities of a different kind. Let us now inquire what difference it would make if there were such other quantities.

First, let us assume that the synoptic principle is true and that all physical observables can be summarized in a ψ-function. In that case, any direct verification of interphenomena is excluded, and there is no way of discriminating between the wave interpretation and the particle interpretation. These interpretations thus constitute *equivalent descriptions*; both are admissible, and they say the same thing, merely using different languages. As a consequence, the causal anomalies play a rather harmless part. They do not affect the observables; and as far as the unobservables, or interphenomena, are concerned, the existence of the anomalies is, so to speak, not localizable. By this term I mean to say that, by the use of a suitable interpretation, the anomalies can always be eliminated from a specific problem, although they will then reappear in some other place. By pushing the anomalies away, the problem under consideration can always be studied on the

---

[5]*PhF*, pp. 26–33. The proof of the principle of anomaly is given in the same work, on pp. 122–129. For a simplified presentation of the principle and the discussion of extension rules, see Hans Reichenbach, *The Rise of Scientific Philosophy* (Berkeley, Calif., Univ. Calif. Press, 1951), chap. xi.

assumption of normal action by contact. When we then turn to investigate a problem for which anomalies have thus been introduced, we can once more switch the interpretation and study this problem, too, in a form free from anomalies. The eliminability of causal anomalies represents an important feature of a quantum mechanics for which the synoptic principle is true; and the switching of interpretations, widely practiced by physicists, is a permissible and legitimate logical device, because wave and particle interpretations constitute equivalent descriptions, each of which is as true as the other.

Second, let us assume that the synoptic principle is false. This would imply that there are observables, though perhaps unknown today, which cannot be included in the ψ-function and the distributions of which are not subject to Heisenberg's indeterminacy relation (1). By the use of such observables, it would be possible to acquire a better knowledge of interphenomena; we could find out whether there are localized occurrences between the phenomena, or whether these occurrences are spread over a wide area. In other words, the new observables would enable us to discriminate between the particle interpretation and the wave interpretation.

Would the new observables allow us to construct strict causal laws instead of mere probability implications? This question must be left open; we have no knowledge of what these new observables would allow us to predict. But let us assume, for a moment, that they would indeed afford to us the possibility of strict predictions, that they would permit us to formulate strict causal laws. What kind of laws would these be? We would then face the fact that the causal anomalies have now become properties of observables. These anomalies could no longer be eliminated, and they would enter into the formulation of causal laws. For instance, if it is assumed that position and momentum of a particle passing through a slit could be measured simultaneously, the conditional causal law would not have the form: If the particle has the position q_1 and the momentum p_1 at time t_1, then the corresponding values at time t_2 would be q_2 and p_2. Instead, it would have the form: If the particle has the position q_1 and the momentum p_1 at the time t_1, and if some other slit is closed, then the corresponding values at time t_2 would be q_2 and p_2; but if the other slit is open, these values would be q_2' and p_2'. The conditional causal law would thus include the expression of an action at a distance. There would be further difficulties; it can be shown that the statistical distribution of other particles would enter as a causal factor into the law, even if those other particles begin their motion at a time when the particle considered has already finished its motion.[6] Other such consequences, not less surprising, will be studied in the following section.

Summarizing, we may say that the existence of observables which cannot be incorporated into the ψ-function and are not subject to the indeterminacy relation—provided such existence leads to strict causal laws at all—would not remove the causal anomalies occurring for those quantities which are included in the ψ-function. Instead, in such a world, the causal anomalies would have entered into the realm of observables. If the synoptic principle were false and a determinism were superimposed on quantum mechanics, it would not be the determinism of classical physics; it would be a determinism in terms of action at a distance and causal anomalies.

The synoptic principle, if it is true, acts as a screen protecting us from a determinism of this kind. Speaking as a physicist, I would therefore say: let us be glad that we cannot observe interphenomena and cannot reintroduce determinism, because what we would thus get would be a very unsatisfactory brand of determinism. Speaking as a philosopher, I would adopt a less emotional attitude toward the issue; but I would admit that a determinism full of causal anomalies is not very likely to exist, in view of all the results of physics, and that therefore such considerations supply strong evidence against the supposition that there are physical quantities not controlled by the ψ-function.

This is, in fact, a conclusion which we can draw: If the synoptic principle were false, the consequences of the known laws of quantum mechanics without the synoptic principle would be so absurd, so unlikely as compared with the known properties of the physical world, that we can draw an inference in reverse and conclude that the synoptic principle is very likely true. As in the illustration of the principle of the limiting character of the velocity of light, the theory as a whole yields positive evidence for the principle. And with the confirmation of the synoptic principle, the indeterminism of quantum mechanics is well established, not merely as a product of absence of knowledge to the contrary, but as the outcome of a consistent theory which lends strong inductive support to the thesis that there is no determinism behind the limitations of predictions. The synoptic principle and the indeterminism of quantum mechanics may therefore be regarded as representing a physical law.

For these reasons, we may say that quantum mechanics is *essentially* a statistical discipline; it is not, like the statistical version of classical thermodynamics, a statistical discipline *faute de mieux*. Classical

---

[6]See *PhF*, §§7 and 8.

physics supplies grounds for believing that, if we could make observations of elementary particles, we could predict the future in more detail and with greater precision. And although classical statistics does not presuppose a determinism of molecular processes, it is quite compatible with such a determinism. It was shown in §9 that the statistical laws of thermodynamics are derivable from the laws of classical mechanics; the canonical equations of motion, by way of Liouville's theorem, supply the probability metric of the phase space, and by way of Von Neumann's and Birkhoff's theorems, also entail the ergodic hypothesis. When we said that the nonuniform convergence of schema (11, 4) does not yield much evidence for determinism, we did not argue that it is incompatible with determinism; it makes determinism an empty addition. It is different for quantum physics. The statistical relationships of quantum mechanics are of such a kind that, if they were combined with a determinism of elementary occurrences, absurd consequences would result; they are inductively incompatible with determinism.

Those who would like to regard quantum mechanics as a provisional solution should have this result in mind. And they should not construct false analogies by referring to the nineteenth-century development of atomic theory. It is true that in the beginning a knowledge of atoms was founded on statistical observations alone and could only later be based on the observation of direct effects of individual particles. But a parallel with quantum physics cannot be drawn. The statistical form of quantum mechanics cannot be regarded as the prelude to a later individualistic treatment of quantum-mechanical systems, because the assumption underlying such a treatment is inductively incompatible with quantum statistics. The ψ-function of quantum mechanics cannot be reconciled with a deterministic background.

Let us discuss this judgment upon determinism in greater logical detail. We do not say that it is meaningless to speak of a determinism behind the indeterminacy of quantum mechanics; we rather maintain that such a statement would be false. We have, of course, only inductive evidence for its falsehood, not a demonstrative proof. But since we can test the statement by an inductive argument, the statement is not meaningless.

The statement referred to in this argument has the form of a conjunction, reading: There are observable physical quantities that cannot be incorporated into the ψ-function and in terms of which the future values of all physical quantities could be strictly predicted. This statement, as explained, is not meaningless, but in all probability false. We can also consider another statement, which has the form of an implication: If we could observe physical quantities that are not subject to the limitations ensuing from the ψ-function, then we could predict strictly the future values of all physical quantities. This implication expresses a conditional contrary to fact in that its antecedent clause is not true; but if the synoptic principle is accepted as a physical law, this implication is meaningless. This is not to say that all conditionals contrary to fact are meaningless. But in order to constitute an admissible conditional contrary to fact, an implication has to satisfy certain conditions, which are laid down in the theory of nomological statements. The implication described violates these conditions, since it is supposed to describe a law of nature, while its antecedent clause contradicts a law of nature. It is therefore physically impossible to verify the implication inductively by observation of cases in which the antecedent clause is true. An implication of this kind uses the word "if" in an illegitimate sense, which does not allow for a counterfactual meaning.[7]

Returning to the question, raised above (p. 213), whether causality in a conditional form is still compatible with the indeterminacy relation, we arrive at the result that causal statements of this form would be meaningless. The statement, "If the physical parameters of the situation at the time t_1 have the values u_1, v_1, . . . , then they will have at a later time t_2 the values u_2, v_2, . . . ", must be interpreted as equivalent to a statement of the form, "If we could observe at time t_1 all the parameters exactly and found the values u_1, v_1, . . . , then we would observe at time t_2 the values u_2, v_2, . . . ". But this is the if-then statement just discussed; therefore the conditional form of causality is meaningless.

Let us compare this result with the discussion of certain if-then statements on page 219. These if-then statements are meaningful only

---

[7]For this problem, see Hans Reichenbach, *Nomological Statements and Admissible Operations* (Amsterdam, North-Holland Publ. Co., 1954), chap. v. If we denote the conjunction described above by "*a.b*", then "*a.b*" is false of second order, in the terminology developed in *Nomological Statements . . .* , according to which terminology a physical law is true of second order. Now if "*a* ⊃ *b*" is not true of the third order, or analytic, then its *T*-case "*a.b*" is false of the same order as that of which the implication is true, and the implication is thus not admissible. An admissible implication is illustrated by the conditional contrary to fact: if the man had touched the high-voltage wire, he would be dead. This implication is true of second order, whereas its antecedent clause is false of first order. In contrast, an inadmissible implication is exemplified by the conditional contrary to fact: if an ice cube of $86°$ C. is heated, it contracts. Here the antecedent clause is false of the second order, whereas the implication is not true of third order, since it is derived from the law of nature that ice, when heated, contracts. This implication is meaningless. In fact, referring to the properties of ice, we can just as well derive the implication: if an ice cube of $86°$ C. is heated, it expands.

because they are made on the assumption that a strict prediction of future values of the parameters is possible; the statements thus have *logical meaning*. But they have no *physical meaning*, since the assumption of strict predictability is false in terms of a physical law. When we say that the conditional form of causality is meaningless, we refer to physical meaning; that is, such supposed laws are *physically meaningless*.[8]

With these considerations, we arrive at an objective meaning of indeterminacy. Heisenberg's relation does not express merely a limitation of human capacities of knowing; it formulates, rather, a physical law holding for all physical quantities.

In many presentations of quantum mechanics, the view has been expressed that the uncertainty derives from the disturbance of the object by the human observer, whose intervention changes the observed quantity in an unforeseeable way. Such conceptions have even been put forward with the claim that quantum mechanics defies an empiricist interpretation and requires a return to idealistic philosophies, according to which the ego creates the world, or the world at least could not exist without an observing ego. I do not think that such views are tenable. The human observer is irrelevant to the indeterminacy; it is not he, or his act of observation, which creates the uncertainty. The indeterminacy is a purely physical affair and can be stated as an objective property of the physical world, without reference to an observer.

First of all, not every observation must disturb. Instruments of measurement, including those applied in quantum physics, can be so constructed that they present their indications as figures on a dial. The reading of these figures by a human observer does not disturb the object. Now it is true that when we draw the line between object and observer in such a way that the instrument of measurement belongs to the observer's side, the object is disturbed; but we need not draw the line in this way. When we include the instrument on the side of the object, the line of demarcation is drawn in such a way that the act of observation does not disturb. It is therefore not the human observer who creates the uncertainty. The indeterminacy, which for an experimental arrangement of this kind still exists, appears, rather, as a relationship between the measuring instruments and quantum phenomena, and thus as a relation between physical objects alone.

Since measuring instruments are macrocosmic objects, we may say that the indeterminacy arises when relationships between macrocosm and microcosm are involved. Heisenberg's principle states that there

---

[8]For the terms *logical meaning* and *physical meaning*, see Hans Reichenbach, *Experience and Prediction* (Chicago, Univ. Chicago Press, 1938), p. 40.

is no way of determining microcosmic quantities in terms of macrocosmic quantities to a higher degree of exactness than that formulated
in the inequality (1). This statement expresses a physical law, a
relation between macrocosm and microcosm. Since human beings are
themselves macrocosmic objects and their perceptual organs respond
only to stimuli in the macrocosmic sphere, their inferences concerning
the microcosm must be based on macrocosmic observations; this is
the reason that human knowledge of the microcosm is limited by the
uncertainty principle. We thus arrive at a very empirical interpretation
of the limitation of human knowledge with respect to quantum phenomena. There is no need for a relapse into idealistic philosophies, if
quantum mechanics is studied under the magnifying glass of logical
analysis.

• 26. The Genidentity of Quantum Particles

The relation of genidentity, which connects different states of the
same thing at different times, was studied in §5. A thing is a series
of events succeeding one another in time; any two events of this
series are genidentical. The conception of physical identity, of an
individual thing that remains the same throughout a stretch of time, is
based on the properties of this relation.

Speaking of things and speaking of events represent merely different
modes of speech. When we wish to translate one of these modes into
the other, we must be careful not to use a mixed language. The sentence of the *thing language*, "This tree is old", must be translated
into *event language* not in the form, "All events constituting this tree
are old", but in the form, "The first events of the series constituting
this tree are separated by a long time stretch from the present event".
Similarly, the sentence, "Peter is taller by two inches than he was a
year ago", is translated as "The present event of the series constituting Peter occupies a spatial volume which is two inches longer than
a corresponding volume was a year ago". If the translation is carefully
done, there will be no difficulties, and the many pseudo-problems are
avoided which philosophers have taken pains to discover in the existence of things that change and yet persist in time.[1]

However, we must investigate the problem of defining the kind of
genidentity which allows us to regard a series of events as states of
a thing. We saw that genidentity is restricted to events which are

---

[1]For a discussion of the two languages, see Hans Reichenbach, *Elements
of Symbolic Logic* (New York, Macmillan Co., 1947), §48.

members of a causal chain. This condition, obviously, is not sufficient to define genidentity. An event may be situated on several causal
chains, which intersect at the event; and the question arises how to
select the chain of physical identity.

For macroscopic objects, we define *material genidentity* in terms
of several characteristics, which can be divided into three groups.
First, we associate material genidentity with a certain *continuity* of
change. When a billiard ball moves, there is a continuous transition
from position to position. If it hits another billiard ball, we would
not accept an interpretation according to which the two balls exchange their physical identity, because such exchange would require
the jumping of one ball into the position of the other. This jumping
would be discontinuous; the center of one ball would jump into the
center of the other. Obviously, such a discrimination is possible only
because our spatial observation is exact enough to distinguish between positions of the ball the distance between which is small compared with the diameter of the ball.

Second, it is a characteristic of material objects that the space
occupied by one cannot be occupied by another. Two billiard balls
occupy different volumes of space. There may arise ambiguities; a
nail driven into a piece of wood may be said to be inside the space
occupied by the wood. But we can redefine the space filled by the
wood in such a way as to exclude the cavity filled by the nail. Thus
spatial exclusion can always be regarded as a necessary characteristic of solid material objects.

Third, we find that whenever two material objects exchange their
spatial positions this fact is noticeable. We usually recognize this
change of position by the use of specific marks on the objects. Billiard balls are distinguishable by their color, or by little scratches
on their surfaces, or by a very precise measurement of their weights,
and so on. These marks remain on the object in accordance with the·
continuity criterion, and permit an identification of the objects even
when no observation during the change of spatial positions was made
and the continuity criterion cannot be applied. In other words, an
interchange of spatial positions is a verifiable change even though no
records of the act of interchanging are available.

These three characteristics of material genidentity are necessary
criteria, but they are not sufficient. When a house is torn down, the
resultant heap of bricks and beams represents the same accumulation
of matter as the house, but the heap is not called the same thing as
the house. The definition of a house requires certain interrelation-
ships between its parts which are no longer satisfied by the heap.[2]

----

[2]See Hans Reichenbach, *Experience and Prediction* (Chicago, Univ. Chicago
Press, 1938), p. 105.

But the following considerations require only the necessary criteria just given.

Some investigation shows that not all marks are transferred in accordance with continuity. If a rolling billiard ball hits another ball which is at rest, it can impart all its speed to the other ball and come to a full stop. Here the mark consisting in the motion is transferred, contrary to material genidentity. This allows us to construct another chain of genidentity: we say that the kinetic energy travels from one ball to the other, and by this usage of words we single out another physical entity, the energy, whose genidentity follows different rules. This is not a material genidentity; it might be called a *functional genidentity*, a genidentity in a wider sense. Also of this functional kind is the genidentity of water waves, which we distinguish from the material genidentity of the water particles; these water particles move up and down, as indicated by a floating piece of wood, while the wave moves horizontally.

Things of a mere functional genidentity may readily violate the second and third characteristics, whereas the first is usually adhered to. Two quantities of energy can exist in the same place; energy satisfies the principle of superposition. And energy cannot be earmarked. When two billiard balls of equal speeds enter into a head-on collision, they return the same way with the same speeds. Were their energies exchanged, or does each ball continue to possess its original energy, while merely its velocity is reversed? This is a meaningless question, because there is no way of identifying energy. It should be noted that here the continuity criterion, though it is satisfied, does not supply an answer either, because both interpretations lead to a continuous genidentity. We have here a situation in which two world lines intersect and we cannot tell which line is which after the intersection. We might try to find a way out of this ambiguity by considering two balls of different speeds. It is true that if one ball traveled originally at a higher speed, this higher speed will be found, after the collision, in the other ball. But we need not conclude that the energies were completely exchanged; perhaps only the surplus energy traveled into the other ball. Such ambiguities are typical for mere functional genidentity. They create no difficulties if it is recognized that genidentity can here be defined in various ways.

Is there some such thing as a "true" genidentity? This question appears nonsensical. We can define "genidentity" to suit our purposes. However, we can distinguish between different kinds of genidentity; and it makes sense to distinguish the material genidentity of the billiard balls from the functional genidentity of the traveling

kinetic energy, or the material genidentity of the water particles from
the functional genidentity of the waves. The reason is that material
genidentity is definable in terms of the three characteristics mentioned—continuity of change, spatial exclusion, and noticeable difference when macrocosmic objects exchange positions.

On closer inspection, we discover that what we call material genidentity is often, in fact, a functional genidentity. The human body
exchanges its material substance continually; after seven years, it
is said, no part of it is materially the same as before. As a consequence, the human body satisfies the continuity criterion only approximately; a precise examination would show that parts of the body
separate from it. But the amount of matter that is exchanged at any
moment is small as compared with the quantity of matter which is
not affected by change during that moment. When a sailboat is repaired
repeatedly and finally every one of its boards and supports has been
exchanged, one can still quite meaningfully call it the same boat,
though again the continuity criterion is only approximately satisfied.
A river replaces its water all the time and thus is functionally rather
than materially genidentical. When Heraclitus realized this fact, he
concluded that change contradicts the conditions of a "true" genidentity and that one cannot step twice into the same river. He should
have concluded instead that functional genidentity is a legitimate
concept and that one can step twice into the same river.

Even when we accept the latter conclusion, however, we should
like to insist that a material genidentity can always be carried through,
if we select the genidentity suitably. If a living body exchanges its
substance all the time, the parts of matter which enter into it, stay
inside it, and then leave it, seem to define lines of material genidentity. And if these parts of matter disintegrate chemically and
divide into other substances, which in turn may be divided by cor-
rosion, by chemical processes, or in other ways, it appears to be a
legitimate statement that the elementary particles of which the parts
of matter are composed define strict lines of material genidentity.
From this viewpoint the theory of atoms was constructed. If macrocosmic genidentity is functional throughout, then there is a material
genidentity behind it in the microcosm; the atoms are those last units
which in their immutable sameness draw lines of material genidentity
through the physical world. This conception stood in the background
of the atomic theories of the ancients; and although modern atoms
are no stable lumps of matter any more, they are supposed to have
elementary constituents ready to play the part of ultimate physical
entities attributed to atoms by Democritus.

In particular, atoms help us understand the material nature of such things as gases and liquids. Two gases mixed and filling the same container violate the second characteristic of material genidentity, since they occupy the same space. By breaking down the gases into very small parts, into atoms, or molecules, their genidentity is reduced to the material genidentity of the atoms. The genidentity character-istics of the solid macroscopic body are thus construed as properties of the ultimate particles of matter, and the liquid and gaseous states of matter are interpreted as reducible to the solid state of atoms.

The question arises in what sense one can speak of the material genidentity of elementary particles. The usual methods of identifi-cation break down in the atomic domain. We cannot observationally test the continuity of motion, though there remained the hope for classical physics that some day we might be able to do so, a sup-position which is made physically meaningless, for quantum physics, by the indeterminacy relation. Reference to the second characteristic, spatial exclusiveness, cannot be made, because there exists no crucial experiment deciding whether the interphenomena consist of waves or particles; if they are waves, they satisfy the principle of superposition. And the presence of the third characteristic cannot be observed either, at least not in the way in which it is observed for macroscopic objects. We cannot make marks on atoms which they keep during their travels, as the zoölogist labels birds to trace them during their migration. It is true, we can observe traces of individual particles in Wilson's cloud chamber; and the physiologist uses radio-active atoms as tracers for the spreading of chemical substances within a living organism. But the question is whether we thus trace a material or merely a functional genidentity. How can we distinguish between the two, if the macroscopic methods of identification are inapplicable?

A good answer to this question was given in classical thermo-dynamic statistics. It suggests that we have to replace an individual examination of particles by inferences based on statistical properties of an assemblage of particles. This program can be made clear by the use of the statistical methods discussed in §8, where we explained the computation of probabilities concerning various distributions of molecules. There we considered arrangements of specific, individual particles situated in different cells, or compartments, either of physi-cal space or of some parameter space. We recall that a distribution is a class of arrangements and that counting arrangements depends on what we understand by *different* arrangements. It was emphasized in §8 that, in classical physics, two arrangements are regarded as different when they result from one another by an interchange of molecules between two given compartments. For instance, if we have two particles, which we name a_1 and a_2, and wish to arrange them in three compartments, this can be done in nine different ways (see table 1). The third characteristic of material genidentity, the requirement that exchange of spatial position should lead to an observable

TABLE 1
MAXWELL-BOLTZMANN STATISTICS
(The particles are distinguishable.)

	Arrangement								
	1	2	3	4	5	6	7	8	9
Compartment 1	a_1 a_2	-	-	a_1	a_2	-	-	a_1	a_2
Compartment 2	-	a_1 a_2	-	a_2	a_1	a_1	a_2	-	-
Compartment 3	-	-	a_1 a_2	-	-	a_2	a_1	a_2	a_1

difference, is thus translated into a statistical property; it supplies the definition of "arrangement".

This way of counting is characteristic for Maxwell-Boltzmann statistics. It was shown in (8, 3) that, if we have n particles and m compartments, or states, the number of possible arrangements is given by

$$\text{Maxwell-Boltzmann: } N(n, m) = m^n . \tag{1}$$

In a similar way, other statistical results are found; for instance, formula (8, 5) is based on the given definition of arrangement. Since this formula, in turn, enters into the value of the entropy, and the laws of entropy are accessible to experimental test, experiments of this kind constitute a test for the adequacy of the given definition of "arrangement" and thus for the assumption of material genidentity of the molecules. The same applies to other physical consequences derived from the definition of "arrangement" in terms of computed probabilities.

Whereas these tests have been positive in the realm of occurrences studied by classical physics, they have turned out negative whenever the quantum nature of elementary occurrences can no longer be neglected—for example, when gases are subjected to very low temperatures. Let us first remark that, for quantum phenomena, the compartments of particle statistics have a peculiar physical significance. They are given by the various quantum states in which a particle can be and which are represented by the *eigen-values* resulting from solutions of Schrödinger's equation.[3] For instance, the possible orbits of the electrons in Bohr's atom model represent such states. The energy of gas particles, likewise, is divided into discrete states. For gas statistics, quantum states find a natural interpretation as follows. In §10 we discussed the phase space built up by the $2un$ coördinates resulting from the position and momentum parameters $q_1 \ldots q_u$, $p_1 \ldots p_u$ of n particles. The product $q_i p_i$ of corresponding parameters has here the dimensions ml^2t^{-1} (m = mass, l = length, t = time), which is the dimension of action, that is, the product of energy and time. This is precisely the dimension of Planck's quantum of action, h; thus the phase space is divided naturally into cells of the size h^{un}, which represent the compartments of quantum statistics. These quantum cells take over the function of the cells into which Boltzmann divides the phase space; in contrast to the cells of the phase space, they do not result from an arbitrary division of space, but are given, so to speak, in an absolute way. They are defined by the fundamental rules of quantum physics.

When the statistics of quantum processes are computed in terms of quantum states, it turns out that the usual Maxwell-Boltzmann methods do not lead to correspondence with observational results. It was shown by Bose and Einstein that the deviation springs from the way of counting arrangements, and that in order to arrive at statistical results conforming to experimental data, we must give up regarding two arrangements as different when the particles in different compartments are interchanged. This means that we must regard the particles as indistinguishable, and hence as violating the third characteristic of material genidentity. The computation in table 1 is then to be replaced by the one given in table 2, in which we have omitted the subscript at the letter "a", and thus have indicated that the particles cannot be distinguished. Here we have only six different arrangements for two particles in three compartments.

The extension of this way of counting to n particles and m states, or compartments, is given as follows. Let us arrange the m compartments all in a row; they then have $m-1$ inner dividing walls, or partitions. When we now arrange the n particles, in some chosen fashion, in the m compartments, we will put the particles within the same compartment in a row. The n particles and the $m-1$ partitions constitute a row of $n+m-1$ things. The number of their permutations is $(n+m-1)!$. Since interchange of particles, or interchange of partitions, leaves the arrangement identical, the number must be divided by

[3]See, for instance, *PhF*, p. 73.

TABLE 2
BOSE-EINSTEIN STATISTICS
(The particles are indistinguishable.)

	Arrangement					
	1	2	3	4	5	6
Compartment 1	a a	-	-	a	-	a
Compartment 2	-	a a	-	a	a	-
Compartment 3	-	-	a a	-	a	a

$n!(m-1)!$. For instance, when we consider the first arrangement, that is, the first column of table 2, we can permute the two particles and the two inner partitions. We then obtain, not only all the other arrangements, but each in four replicas, the number 4 resulting as the product $2!2!$. We thus find:

$$\text{Bose-Einstein: } N(n, m) = \binom{n+m-1}{n}. \tag{2}$$

This way of counting enters into all kinds of computations of thermodynamic quantities. The results have been well confirmed by experiments. Among such observational confirmations, the computation of the chemical constant of a gas may be mentioned; furthermore, liquid helium obeys Bose statistics, as is seen from the strange behavior of its specific heat, the mathematical theory of which was given by F. London.

There are other physical processes which require a third form of statistics. Here, too, the particles are indistinguishable; but in addition, they are subject to an exclusion principle first discovered by W. Pauli and extended to gas statistics by E. Fermi and P. A. M. Dirac. There cannot be more than one particle in a given state. This kind of statistics is presented in table 3. We have here only three different arrangements for the two particles in three compartments.

For n particles and m states, or compartments, we arrive at the result:

$$\text{Fermi-Dirac: } N(n, m) = \binom{m}{n}. \tag{3}$$

This formula is found as follows. For one particle, we have m possible

TABLE 3
FERMI-DIRAC STATISTICS
(The particles are indistinguishable and
subject to an exclusion principle.)

	Arrangement		
	1	2	3
Compartment 1	a	-	a
Compartment 2	a	a	-
Compartment 3	-	a	a

arrangements. When we add another particle, there are only $m-1$ compartments free for it, because of the exclusion principle; thus we have $m(m-1)$ arrangements. This must be divided by 2! because interchanging the particles leaves the arrangement unchanged. The method is easily extended to n particles.

Again, the values of $N(n, m)$ enter into various kinds of statistics. It applies to certain groups of physical phenomena—for example, to the "gas" of conduction electrons within a wire through which an electric current flows.

That the nonclassical statistical formulas (2) and (3) yield results that correspond to observational facts cannot be questioned. The question is what these facts prove with respect to the problem of genidentity. It is usually agreed that they prove that the particles are indistinguishable, and we have thus far followed this conception. We must now investigate whether this conception is correct, or subject to qualifications.

We said above that the material genidentity of particles is not accessible to individual examination and that instead of making such an examination we have to draw inferences based on statistical properties of an assemblage of particles. What kind of inferences are we making here? Obviously, we are drawing inferences concerning interphenomena, since genidentity refers to physical identity extending from one observable phenomenon to another. These inferences, therefore, presuppose certain extension rules. Now we cannot ask whether an extension rule is true; we can only ask at what consequences we arrive if we use a certain extension rule. The extension rule assumed for the given interpretation, which is the rule usually employed in physics, is that all distinguishable arrangements must be equally probable; it then follows that, for instance for Bose statistics, only the arrangements presented in table 2 are distinguishable from one another. The question is, whether this is the only possible extension rule, or whether some other rule would reintroduce a physical identity of individual particles.

This question can be answered by the following method. We use as the extension rule not the equiprobability of distinguishable arrangements, but the rule that the particles be considered as individually distinguishable. Since the probability metric at which we must arrive is given, say, by Bose statistics, this rule leads to unequal probabilities for distinguishable arrangements. Let us see what this means.

When we toss two coins with one hand, we have four possible arrangements. If we denote heads by "*H*" and tails by "*T*" and indicate the individual coin by a subscript, these possible arrangements can be written:

$$H_1 H_2 \ , \quad H_1 T_2 \ , \quad T_1 H_2 \ , \quad T_1 T_2 \ . \tag{4}$$

The assumption that these four arrangements are equiprobable is made in classical statistics; and actual observation confirms this assumption when the coins are thrown repeatedly. The frequency of the result "heads only" amounts to 1/4; the frequency of the result "tails only", likewise to 1/4; but that of the result "1 head and 1 tail" amounts to 1/2, because this result includes two possible arrangements.

Now let us suppose that statistics lead to different results and that each of the three cases named occurs in 1/3 of all throws. This is the statistics of *nonindividualized combinations*, that is, Bose statistics. It would be symbolized by the three cases

$$H H \ , \quad H T \ , \quad T T \ , \tag{5}$$

in which we have omitted the subscripts. If frequent observation confirmed these statistics, for which the three arrangements named in (5) would be equiprobable, would we be compelled to say that the coins are indistinguishable?

Such a conclusion would be unjustified. Since we can easily distinguish the coins—by a visible mark, for example—we would say that the four possible arrangements (4) still occur, but that they are not equally probable. The first and the last arrangement of (4) would each have the probability 1/3, and the two middle ones would each have the probability 1/6. On this assumption, the three arrangements of (5) would be equally probable.

This interpretation would mean that the throws of the coins are not independent events. If one coin shows heads, the other would have a tendency to show the same result; and likewise for tails. Denoting the events heads on the first and the second coin by "B_1" and "B_2" respectively, we have here:

$$P(B_1) = \tfrac{1}{2} , \qquad\qquad P(B_2) = \tfrac{1}{2} ,$$

$$P(B_1, B_2) = \tfrac{2}{3} > P(B_2), \quad P(B_2, B_1) = \tfrac{2}{3} > P(B_1) ,$$

$$(6)$$

and correspondingly:

$$P(\overline{B_1}, B_2) = \tfrac{1}{3} < P(B_2), \quad P(\overline{B_2}, B_1) = \tfrac{1}{3} < P(B_1) . \qquad (7)$$

This means that the events would be causally dependent on one another. The sequence of events given by the throws of one coin would be statistically, and thus causally, dependent on the other sequence.

Corresponding relations are easily constructed for a disjunction of n events and two sequences.[4] If three or more sequences are involved, the dependence relations are more complicated. For instance, for three sequences the attributes of which are denoted by "B", "C", and "D", we have:

$$P(B, C) \neq P(C), \qquad\qquad P(C, B) \neq P(B),$$

$$(8)$$

$$P(B \cdot C, D) \neq P(D), \qquad\qquad P(C \cdot D, B) \neq P(B),$$

and similar relations resulting when the letters are interchanged. The quantities with two letters before the comma are here unequal to those with one letter before the comma. This means that one sequence is dependent, not only on each other one, but also on the two others together.[5]

We arrive at the following result. Assume that we could assign material genidentity to each particle of a Bose ensemble; then we would find that the particles are mutually dependent in their motions. If one particle is in a certain state, then there exists a tendency for the others to go into the same state; and even more complicated dependence relations exist for combinations of particles, similar to (8). These causal relationships would represent action at a distance, since the particles can be far apart; that is, the dependence relations would constitute causal anomalies. In other words: any assignment of physical identity to Bose particles leads to causal anomalies.

We see now how the thesis concerning indistinguishable particles is to be qualified. In precise language we cannot simply say: The particles are indistinguishable. We must say: Either the particles are indistinguishable, or their behavior displays causal anomalies. We

---

[4] See *ThP*, p. 156.
[5] See *ibid.*, p. 105.

are left the choice of selecting the one or the other interpretation. Neither interpretation is "more true" than the other; the two are equivalent descriptions.

However, only one of the two descriptions supplies a normal system, that is, a system free from causal anomalies; this is the description according to which the particles are indistinguishable. When we follow the usual rule of employing a normal system whenever it is possible, we may therefore say, without hesitation, that the particles are indistinguishable.

For Fermi statistics, the situation is somewhat different. Fermi statistics does not offer any difficulties to an interpretation in which the particles are distinguishable. For this interpretation, each arrangement of Fermi statistics would be multiplied by the constant factor $n!$, because this is the number of permutations of n particles. The number of all possible arrangements given in (3) is thus multiplied by $n!$; but the equiprobability relations would not be changed. This means that any two Fermi arrangements, which are now classes of individual arrangements, would still be equiprobable.

The causal anomaly of Fermi statistics consists, rather, in the exclusion principle, if that principle is interpreted as a force acting between individual particles. This means that if we assign physical identity to the particles, the exclusion represents a strong mutual dependence of a negative kind. If one particle is in a certain state, then no other particle is in that state.

These considerations supply an important contribution to the problem of determinism. This problem was discussed in §25 as a question of inductive evidence. We derived certain consequences which would follow if the hypothesis of determinism were true, and then turned the unlikely nature of these consequences into inductive evidence against the hypothesis. We can construct a similar argument from the result of our analysis of Bose statistics.

If the synoptic principle of quantum mechanics were false and we could somehow observe position and momentum of a particle simultaneously, we could follow the particle in its path and thus verify its physical identity by means of the continuity characteristic. Since particles are supposed to satisfy the characteristic of spatial exclusiveness, this identification would lead to a material genidentity. Now we saw that for materially genidentical particles Bose statistics leads to causal anomalies. We thus arrive once more at a situation in which a causal anomaly has become a property of observables. This conclusion supplies inductive evidence against the hypothesis of determinism and for the synoptic principle.

Conversely, if we accept the synoptic principle, its combination with Bose statistics leads to the result that there is no material genidentity for elementary particles, provided we describe the physical world in terms of a normal system, that is, a system without causal anomalies. It follows that, in the sense of the term as defined above, there is no material genidentity at all in the physical world; there is only functional genidentity. The conception of material genidentity turns out to be an idealization of the behavior of certain macroscopic objects, the solid bodies, which behavior, however, corresponds only approximately to the idealization. Extending the idealization to atoms in the hope of finding it satisfied in this domain—the program of the classical theory of atoms—does not help. In the atomic domain, material genidentity is completely replaced by functional genidentity.

• 27. *The Entropy Concept of Quantum Statistics*

The assumption that particles are indistinguishable leads to a new measure of entropy, which must now be studied. It will be the aim of our investigation to find out under what conditions the two measures of entropy, the quantum-statistical and the classical concept, lead to observable differences, and under what conditions they lead to indistinguishable results for macroscopic observations.

Classical statistics, as presented in §8, is based on a *particle lattice*, which corresponds to the lattice (12, 6) and may be illustrated by table 4. Each row, or horizontal line, represents the history of an individual particle. Assuming that we have n particles, we have n rows. Dividing time into small intervals Δt, we write down, for each successive moment t_1, t_2, . . . , the momentary state C_i of the particle. The states C_i thus are used as the attributes named in the lattice. Let us assume, for the present, that all states belong to the same energy level; then there are no restrictions to the columns. Two columns are identical if they correspond, place for place, in the attribute C_i; they then represent the same arrangement. The probability metric of Maxwell-Boltzmann statistics is given by the hypothesis that all columns, or arrangements, are equally probable, or equally frequent in horizontal counting.

When we count for each column how often a state C_i occurs, we find the occupation number n_i of this state. A distribution D is given when, for a column, the occupation number is given for each state. Two columns belong to the same distribution if, for every state C_i, the occupation numbers n_i of the two columns are the same. Consequently, if we

TABLE 4

PARTICLE LATTICE

(The rows are the histories of the particles. The attributes are the states C
in which the particles are at successive times $t_1, t_2, \ldots$)

	t_1	t_2	t_3	$\cdots$
Particle 1	C_3	C_1	C_3	$\cdots$
Particle 2	C_1	C_4	C_6	$\cdots$
Particle 3	C_6	C_5	C_1	$\cdots$
....	....	....	....	....
Particle n	C_4	C_1	C_6	$\cdots$

consider two columns for which the occupation numbers n_i and n_k of two states C_i and C_k are exchanged, these two columns belong to different distributions. Exception is to be made for the case that $n_i = n_k$. A distribution D is therefore identical with the ordered set $\{n_i\}$ of the occupation numbers n_i, $i = 1, 2, \ldots m$, belonging to a column. The order of the set is determined when we introduce some suitable ordering rule according to which the subscript i increases for the states C_i.

When the term "distribution" is defined as in the paragraph just above, the number N_D of arrangements belonging to the same distribution is given by (8, 5). Let us repeat here the formulas derived in (8, 3–8) and write down the *characteristic formulas for Maxwell-Boltzmann statistics:*

$$(1) \qquad N_D = \frac{n!}{n_1! \ldots n_m!} \ , \qquad\qquad N = m^n \ , \qquad (2)$$

$$\sum_{i=1}^{m} n_i = n \ , \qquad (3)$$

$$W = \frac{N_D}{N} \ , \qquad (4)$$

$$S = \log W = n \log n - \sum_{i=1}^{m} n_i \log n_i - n \log m \ . \qquad (5)$$

We use here the relative-normalized entropy S which corresponds exactly to the probability W.

If we tried to base Bose-Einstein statistics on the particle lattice, we should find that the metrical probability hypothesis is not satisfied. In whatever way we assume the genidentity of particles, the various column types, each including identical arrangements in the sense defined, would here not be equally frequent. Consequently, we would have to speak of mutual dependence between particles, which fact expresses causal anomalies; this dependence was studied in the discussion of (26, 6–7). For this reason, a particle lattice is inconvenient for quantum statistics.

For quantum statistics, we therefore construct a different lattice which we call a *state lattice* and which is illustrated in table 5. Each row represents the history of a state C_i. We will again assume that there are m states; then we have m rows. As attribute we write down the occupation number n_i of the state at successive moments $t_1, t_2, \ldots$ Since we do not indicate any individuality of the particles by the use of this number, the state lattice lends itself to a treatment of quantum statistics. We do not speak here of the probability that a certain particle will enter into a certain state C_i, a concept which refers to the particle lattice and leads to causal anomalies. We rather speak of the probability that a state C_i will be occupied by r particles.

We will call two columns identical, or say that they belong to the same arrangement, if they have the same occupation number in corresponding places. This means that the states are regarded as distinguishable individuals; an exchange of two different occupation numbers in the same column leads to a new arrangement. Not all imaginable arrangements are possible, because the number of particles is constant; the occupation numbers of each column are thus subject to condition (3). For the present, we will again consider only the situation in which all states belong to the same energy level; then (3) is the only restrictive condition for the columns. Those columns that actually occur in a closed system will, of course, always satisfy this condition. The probability metric of quantum statistics is now given by the hypothesis that all possible columns of the state lattice are equally probable, or equally frequent in a count in the horizontal direction.

It should be noted that what we call here an arrangement is exactly what we call, in Maxwell-Boltzmann statistics, a distribution. The number N_D of Maxwell-Boltzmann statistics is thus replaced by the number 1. In order to obtain statistics and to define a distribution as a class of arrangements, we therefore must introduce a suitable grouping of arrangements. This is achieved as follows. The possible occupation numbers are integers $r = 0, 1, 2, \ldots n$; when the occupation number of a state C_i is known, we write $n_i = r$. Now we can count how many states have a given number. We can count, for instance, how many states have the occupation number $r = 3$; the number of these states will be called m_3. It may happen that there is no state having the occupation number 3; then we have $m_3 = 0$. In this way we ascertain, for each

TABLE 5
STATE LATTICE

(The rows are the histories of the states. The attributes are occupation numbers, that is, the numbers of particles in the state at successive times $t_1, t_2, \ldots$)

	t_1	t_2	t_3	$\cdots$
State 1	4	0	0	$\cdots$
State 2	2	3	4	$\cdots$
State 3	0	1	2	$\cdots$
$\cdots\cdots\cdots\cdots$	$\cdots$	$\cdots$	$\cdots$	$\cdots\cdots\cdots\cdots$
State m	4	0	1	$\cdots$

column, the set of *occupation frequencies* m_r, $r = 0, 1, 2, \ldots n$. A distribution D is given when this set $\{m_r\}$ is known. Thus two columns, or arrangements, belong to the same distribution D if they furnish the same set $\{m_r\}$.

Note that the m_r of the set are ordered only with increasing size of r; this order has nothing to do with the order of the states C_i in the lattice. As a consequence, a distribution can include arrangements which are macroscopically distinguishable. Suppose we construct the states by dividing the container of a gas into a number of cells, and suppose we consider a distribution in which the number of empty cells is rather large. If we assume that all the empty cells are situated on one side of the container, we arrive at an arrangement in which the gas is concentrated on one side of its container; if we assume that the empty cells are distributed at random, the resulting arrangement may resemble a homogeneous distribution. The two arrangements, which are macroscopically distinguishable, each possess the same set $\{m_r\}$. In this respect, the quantum-statistical distribution differs from the classical one, which never includes macroscopically distinguishable arrangements.

We will first apply the new definition of a distribution to Bose-Einstein statistics. Since there are m states, we have m occupation numbers in a column; they can be arranged in $m!$ ways. However, interchange of equal occupation numbers leaves the arrangement unchanged; therefore we have to divide $m!$ by the number of permutations for each group m_r. We thus construct the number N_D in analogy with (1), replacing the "n" by "m". Using this form of computation, we write down the following *characteristic formulas for Bose-Einstein statistics*, using (26, 2):

$$(6) \qquad N_D = \frac{m!}{m_0! \ldots m_n!} , \qquad\qquad N = \binom{n+m-1}{n} , \qquad (7)$$

$$(8) \qquad \sum_{r=0}^{n} m_r = m , \qquad\qquad \sum_{r=1}^{n} r m_r = n , \qquad (9)$$

$$W = \frac{N_D}{N} , \qquad (10)$$

$$S = m \log m - \sum_{r=0}^{n} m_r \log m_r - \log N , \qquad (11)$$

$$\log N = (n + m - 1) \log (n + m - 1) - n \log n - (m - 1) \log (m - 1) . \qquad (12)$$

Relation (9) takes the place of (3); note that it would not make any difference if we begin the summation in (9) with $r = 0$. A new condition is added in (8), expressing the fact that the number of possible states is a constant for all distributions. For (11)–(12) we have used Stirling's approximation for factorials, neglecting the last term in (8, 6).

For Fermi-Dirac statistics, the corresponding relations are greatly simplified. They result when the summation in the preceding formulas is restricted to the values $r = 0$ and $r = 1$. Furthermore, since a compartment can now contain only one particle, we have $m_1 = n$. Making these substitutions in (6), we see that the value of (6) becomes identical with the value of (26, 3); that is, the value N_D is identical with the value N. In fact, we have here only one possible distribution. These relations lead to the following *characteristic formulas for Fermi-Dirac statistics:*

$$(13) \qquad N_D = \frac{m!}{m_0! \, m_1!} = \binom{m}{n} , \qquad N = \binom{m}{n} , \qquad (14)$$

$$(15) \qquad m_0 + m_1 = m , \qquad\qquad m_1 = n , \qquad (16)$$

$$W = \frac{N_D}{N} = 1 , \qquad (17)$$

$$S = m \log m - m_0 \log m_0 - m_1 \log m_1 + \text{const} . \qquad (18)$$

Since $W = 1$, we have here $S = 0$; to avoid this consequence, we have written S with an undetermined constant. We shall later use only the absolute entropy given by $\log N_D$. For Fermi-Dirac statistics, the entropy concept is of practical use only when different energy levels are considered, as will be seen in §28.

The advantage of the state lattice of quantum statistics is that it permits the use of classical statistical methods, inasmuch as its columns are equiprobable and its probabilities can be computed by counting columns. It is true that the rows of this lattice are not mutually independent, because they are linked together by the restrictive condition (9), which can also be written in the form (3). When we write the occupation numbers in the form of a matrix n^{ki}, relation (3) can be transcribed into the condition, applying to every column i:

$$\sum_{k=1}^{m} n^{ki} = n \ . \tag{19}$$

This condition expresses a statistical dependence. For instance, if the n^{ki} of the first $m-1$ places of a column i add up to $n-1$, the value n^{mi} at the last place must be equal to 1. However, a statistical dependence of this kind does not represent a causal anomaly; it springs merely from the conservation of the number of particles and is thus causally explained. This kind of dependence must be carefully distinguished from the physical dependence holding for the rows of a *particle* lattice controlled by Bose statistics, as discussed with espect to (26, 5-7).

To make this point clearer, let us remark that the statistical dependence of the rows of the state lattice does not originate from some influence of the travel of one particle upon another. This statement can be given the following precise meaning. Suppose that condition (19) is omitted and that the lattice satisfies the two following conditions:

Condition 1. All values $r = 0, 1, \ldots n$ are equiprobable for the n^{ki} in every row k, counting horizontally.

Condition 2. The rows are completely independent of one another.

It is a consequence of these assumptions that all columns are equiprobable. Among the columns there will be some which, by chance, satisfy condition (19). When we now consider the sublattice consisting of those columns that happen to satisfy (19), we find that all its statistical properties can be derived from the two given assumptions, including the equiprobability of its columns and the statistical dependence of its rows on one another.[1]

---

[1]In this derivation we make use of the rule of reduction (*ThP*, p. 97) and regard the sublattice as a disjunction of $\binom{n+m-1}{n}$ possible arrangements satisfying (19). The consideration can also be extended to a lattice having columns that are infinite; we then replace (19) by the condition

Only this sublattice, of course, can be physically realized. In the sense explained, the quantum-mechanical state lattice, which is identical with the sublattice, does not display a causal influence between particles; its statistical properties are derivable from the independence assumption in combination with a restrictive condition for columns. A lattice subject to restrictive conditions for columns can be treated by the methods of classical statistics; in fact, the particle lattice of Maxwell-Boltzmann statistics, if subject to the energy condition (9, 3), is of this type.

The most important question concerns the probabilities w_r of the values $r = 0, 1, \ldots n$ for the n^{ki} in the lattice that is subject to condition (19). Because of this restriction, the w_r do not have the same value for various r. They are defined as follows, when we write "r^{ki}" for the case that $n^{ki} = r$:

$$P\left(r^{ki}\right)^i = w_r \ . \tag{20}$$

If the distribution is stationary and m is large, we can assume that the same probabilities hold for counting along a column; thus we have

$$w_r = P\left(r^{ki}\right)^i = P\left(r^{ki}\right)^k = \frac{m_r}{m} \ . \tag{21}$$

The values w_r can be found by the help of combinatorial methods.

Before entering into this computation, let us insert a remark concerning Maxwell-Boltzmann statistics. It is, of course, also possible to construct a state lattice for Maxwell-Boltzmann statistics. But its columns would not be equally probable. This is seen as follows. What we call an arrangement in quantum statistics corresponds to N_D arrangements of Maxwell-Boltzmann statistics; since in Maxwell-Boltzmann statistics all the N_D arrangements are equally probable, the frequencies of the columns in the state lattice are not equal, but are multiplied by factors N_D, and the values N_D differ for various column-types. The unequal probabilities for the columns make the state lattice inconvenient for Maxwell-Boltzmann statistics. However, the probabilities w_r exist here, too, and govern rows and columns alike; and relation (21) holds likewise for the stationary distribution of Maxwell-Boltzmann statistics.

$$\lim_{m \to \infty} \frac{1}{m} \sum_{k=1}^{m} n^{ki} = q \ . \tag{19a}$$

Requiring a uniform convergence for all columns, we can apply the derivations in this section to the lattice of infinite columns in the form of limit properties.

We will now turn to computing the values of the w_r for the three kinds of statistics. For classical statistics, we have to go back to the metric defined in the particle lattice, which leads to unequal probabilities in the state lattice.

Suppose we assign r particles to one state C_i. Let N_r be the total number of arrangements which agree in that the state C_i has exactly r particles. This number is found by arranging the remaining $n - r$ particles in the remaining $m - 1$ compartments. For Maxwell-Boltzmann statistics, we have to consider, in addition, that there are $\binom{n}{r}$ ways of putting r of the n particles into the compartment C_i. For Fermi-Dirac statistics, r can only have the values 0 and 1. We thus find, using (26, 1-3):

Maxwell-Boltzmann statistics:

$$N_r = \binom{n}{r} \cdot N(n-r, m-1) = \binom{n}{r}(m-1)^{n-r} , \qquad (22)$$

Bose-Einstein statistics: $N_r = N(n-r, m-1) = \binom{n-r+m-2}{n-r}$, (23)

Fermi-Dirac statistics: $\quad \begin{cases} N_0 = N(n, m-1) = \binom{m-1}{n} , \\[2mm] N_1 = N(n-1, m-1) = \binom{m-1}{n-1} . \end{cases} \qquad (24)$

Because of the probability metric assumed for the lattice, the quantity w_r of (20) is now given by:

$$w_r = \frac{N_r}{N} . \qquad (25)$$

When we insert here the values (20)–(22), we will use approximation methods which apply if both m and n are large as compared with r. We employ the following formulas, which hold when n is large as compared with x, or r, respectively:

$$(26) \qquad \left(1 + \frac{x}{n}\right)^n \cong e^x \qquad = \frac{n!}{(n-r)! \, r!} \cong \frac{n^r}{r!} \qquad (27)$$

Calling formula (27) an approximation means that the quotient of the two sides goes to 1 with increasing n. Incidentally, (27) is easily derivable when we write $\binom{n}{r}$ in the form $n(n-1) \ldots (n-r+1)/(r!)$ and neglect the powers lower than n^r. Furthermore, we replace "$m-1$" by "m", and "$n-r$" by "r", if these terms do not occur under n-th powers.

Carrying out these approximations,[2] we find

Maxwell-Boltzmann statistics: $w_r = \dfrac{1}{r!}\, e^{-q}\, q^r$,   (28)

Bose-Einstein statistics: $w_r = \dfrac{1}{1+q} \cdot \left(\dfrac{q}{1+q}\right)^r$,   (29)

Fermi-Dirac statistics: $\begin{cases} w_0 = 1 - q \ , \\[2ex] w_1 = q \ , \end{cases}$ (30)

where, throughout,

$$q = \frac{n}{m} .   (31)$$

The quantity q measures the average number of particles per state. For Fermi-Dirac statistics, we have always $q \leq 1$; for the other two kinds of statistics, we can have $q > 1$.

Concerning formula (28) we are interested, in particular, in the value r_{max} of r, for which w_r is a maximum. Putting here $\dfrac{dw_r}{dr} = 0$, we easily find $r_{max} = q$. This means that the most probable occupation number is given by n/m. An equal distribution of all particles over all states represents, therefore, the most probable distribution, in correspondence with the result (8, 14).

To study formula (29), let

$$\frac{q}{1+q} = w \ , \qquad \frac{1}{1+q} = 1 - w \ .   (32)$$

Then (29) can be written, using (31):

---

[2]The exact values of (28) and (29) can be written in the forms, respectively,

(28a) $w_r = \binom{n}{r} \left(\dfrac{m-1}{m}\right)^n \cdot \dfrac{1}{(m-1)^r}$; $w_r = \dfrac{m-1}{u-r} \cdot \dfrac{\binom{n}{r}}{\binom{u}{r}}$, $u = n + m - 1$.   (29a)

The values (30) are exact.

Bose-Einstein statistics: $\qquad w_r = (1-w)w^r \; , \qquad w = \dfrac{n}{n+m} \; .$ (33)

Letting r equal 0, we find $w_0 = 1-w$. This is the probability of finding an empty state, and also the maximum probability for any occupation number r, since $w < 1$. We thus have $r_{max} = 0$. The empty state is the most probable one, and Bose statistics has the effect that the particles are crowded together in certain states. However, it may be still more probable that a state is occupied by at least one particle than that it is empty. The probability that a state is occupied by at least one particle equals w. This follows because $1-w$ is the probability of an empty state; but it is also consistent with the sum of the geometrical series, when we use the formulas

$$\sum_{r=0}^{\infty} w^r = \frac{1}{1-w} \; , \qquad \sum_{r=1}^{\infty} w^r = \frac{w}{1-w} \; , \qquad (34)$$

and regard n as infinite for the summation of all possible values w_r.

For Fermi-Dirac statistics, the only possible occupation number of a nonempty state equals 1. Therefore q is here the probability that a state is occupied. Whether or not being occupied is more probable than being empty, and whether r_{max} is equal to zero or to 1, depends on whether q is greater than or less than $1/2$.

When we compare the three formulas (28)–(30), we find a peculiar difference. The maximum w_r of Maxwell-Boltzmann statistics, computed from (28), is so constructed that all states, or compartments, can very well be occupied simultaneously by the same $r_{max} = n/m$; equal distribution of all particles over all compartments is thus the most probable distribution. For the two other kinds of statistics, however, it is impossible that all compartments have the occupation number r_{max} at the same time. For Bose-Einstein statistics, in which $r_{max} = 0$, it is impossible that all compartments are empty, unless there be no particles at all. And for Fermi-Dirac statistics neither the value $r_{max} = 0$ nor the value $r_{max} = 1$ can hold for all compartments, unless $n = 0$ or $n = m$, which are exceptional cases. Therefore, the statistical equilibrium of quantum statistics is not given by equal distribution of all particles over all compartments, but by a characteristic variation of occupation numbers, the distribution of which is governed by the probabilities w_r. Although the states change their occupation number with time, the occupation frequencies remain the same, in the stationary distribution.

Could we ever count occupation numbers of states directly, we should find noticeable differences among the three kinds of statistics. However, it can be shown that the differences decrease when q, the average number of particles per state, decreases. If $q \ll 1$, we find, replacing "$1 + q$" by "1" and "e^{-q}" by "$1/(1+q)$" in (28) and (29):

$$\frac{w_r(B - E)}{w_r(M - B)} \sim r! \tag{35}$$

Now we have $r! = 1$ for $r = 0$ and $r = 1$. It follows that, if q is small, the two occupation numbers 0 and 1 have virtually equal frequencies for Bose-Einstein and for Maxwell-Boltzmann statistics.

A similar equality results for a comparison of Maxwell-Boltzmann statistics and Fermi-Dirac statistics. When q is small and we put "$1 - q$" for "e^{-q}" in (28), we find for $r = 0$ and $r = 1$:

$$\frac{w_0(F - D)}{w_0(M - B)} \sim 1 \; , \qquad \frac{w_1(F - D)}{w_1(M - B)} \sim 1 \; . \tag{36}$$

Only for occupation numbers $r > 1$ do differences result for the three kinds of statistics. But if q is small, the frequencies of such numbers are governed by the powers q^2, q^3, ... and are low as compared with the frequencies of the occupation numbers 0 and 1. The exclusion principle of Fermi-Dirac statistics, in particular, is virtually always satisfied for low probabilities of independent events, since q^2, q^3, and so on, can then be neglected as compared with q.

The three kinds of statistics may be illustrated by a numerical example, presented in table 6, for which we have chosen small numbers n and m. The values given for N and N_r are precise values computed from (2), (7), (14), and (22)–(24). The value w_r is computed from (25). The left-hand part of the table is computed for $n = 2$ and $m = 3$, which are the values used for tables 1–3, presented in §25. Here the differences in the values w_r are plainly visible. The right-hand part of table 6 is computed for the same value $n = 2$, but for the larger value $m = 10$. Thus, going from the left-hand part of the table to the right-hand part corresponds to increasing the number of possible states while leaving the number of particles unchanged. The figures in the right-hand part of the table show that here the values of w_r are virtually the same for the three kinds of statistics.

We arrive at the result that the transition to small q, and thus the diluting of the gas with respect to its possible states, leads to a mutual convergence of the three kinds of statistics. The physical interpretation of this result may be postponed until we have studied the influence of a variation in energy levels on the distribution (§28).

Comparing the entropy concept (5) of classical statistics with the

TABLE 6
COMPARISON OF MAXWELL-BOLTZMANN (M-B), BOSE-EINSTEIN (B-E), AND
FERMI-DIRAC (F-D) STATISTICS

Statistics	$n = 2$ $m = 3$						$n = 2$ $m = 10$					
	N	N_0	w_0	N_1	w_1	N_2	w_2	N	N_0	w_0	N_1	w_1	N_2	w_2
M-B	9	4	0.44	4	0.44	1	0.11	100	81	0.81	18	0.18	1	0.01
B-E	6	3	0.50	2	0.33	1	0.13	55	45	0.818	9	0.164	1	0.02
F-D	3	1	0.33	2	0.67	0	0	45	36	0.80	9	0.20	0	0

LEGEND: n = total number of particles
m = number of states, or compartments
N = total number of possible arrangements
N_r = number of arrangements for which a specified compartment is filled by r particles
w_r = probability of finding r particles in a specified compartment = N_r/N.

entropy concepts (11) and (18) of quantum statistics, we notice some important differences. It has been mentioned earlier in this section that what we call an arrangement in quantum statistics corresponds to a distribution of Maxwell-Boltzmann statistics; in the latter statistics, a quantum arrangement would therefore be multiplied by the factor N_D given in (1). Using the occupation frequencies m_r, we can transcribe (1) into the form

$$\text{Maxwell-Boltzmann:} \quad N_D = \frac{n!}{(0!)^{m_0} (1!)^{m_1} \ldots (r!)^{m_r} \ldots (n!)^{m_n}} . \quad (37)$$

We see that all those quantum arrangements which belong to the same quantum distribution, given by a set $\{m_r\}$, are multiplied by the same factor, namely, by the value N_D of (37). For different quantum distributions, that is, different sets $\{m_r\}$, however, the factor changes its value.

If we were to replace the entropy forms (11) and (18) of quantum statistics by the entropy form of Maxwell-Boltzmann statistics, we would have to add the value $\log N_D$, computed from (37), to the forms (11) and (18); and since (37) varies with the set $\{m_r\}$, the maximization properties of the resulting entropy expression would be different. This is the reason that the two kinds of quantum statistics lead to different forms of the stationary distribution, that is, of the distribution associated with maximum probability. Apart from this difference concerning the statistical structure of the state of equilibrium, however, the two kinds of entropy are of the same logical nature. They are functions characterizing the state of a macrocosmic system, functions the mathematical form of which is derived from probabilities and which interpret the process of evolution toward an equilibrium as a tendency to go from states of lower to states of higher probability.

There is, however, a difference which makes the quantum-statistical form of entropy superior to the classical one. In §8 we discussed Gibbs's paradox and showed that an absolute entropy cannot be satisfactorily normalized in classical statistics. It is different with the quantum-statistical entropy; here Gibbs's paradox is automatically eliminated by the mathematical form of the entropy. We recall that this paradox refers to the total entropy of two containers, filled with gases in equal conditions, which entropy should not change when a window between the containers is opened, and should be given by the sum of the individual entropies. This problem was discussed with respect to (8, 17). For Bose statistics, the problem is solved as follows. If in the entropy equation (11) of Bose-Einstein statistics we replace "m" by "$2m$" and "n" by "$2n$", the occupation frequencies m_r are likewise to be doubled. We thus find, omitting the constant "$-\log N$",

$$
\begin{aligned}
S' &= 2m \log 2m - \sum_{r=1}^{2n} 2m_r \log\left(2m_r\right) \\
&= 2m \log m + 2m \log 2 - 2\sum_{r=1}^{2n} m_r \log m_r - 2m \log 2 = 2S \ .
\end{aligned}
\tag{38}
$$

Here we neglect the difference resulting from extension of the summation to $2n$ instead of n, because high values of r do not contribute very much to the sum. Writing "$2S$" on the right would also be correct if the entropy (11) includes an additive constant, if only this constant, as for Maxwell-Boltzmann statistics, is of the form "$n \cdot \text{constant}$". We see that the term "$2m \log 2$" is here eliminated by the combination of the first and second terms, in contrast to (8, 17), where the corresponding term "$2n \log 2$" does not drop out unless the third term in (8, 8) is used. We thus find that the *additivity postulate* is satisfied for the absolute entropy of Bose-Einstein statistics. The same applies to the entropy equation (18) of Fermi-Dirac statistics. For quantum statistics, there exists no paradox of normalization.

The physical explanation is simple. When we open the window between two containers filled by a gas in equal conditions, the number N_b of possible arrangements for the state of equilibrium increases enormously for Maxwell-Boltzmann statistics, because the molecules of one container can now exchange their places with those of the other container. However, if the particles are indistinguishable, such exchange does not count, and the number of arrangements is the same before and after opening the window. On this condition, the absolute number of arrangements can easily be used as a measure of entropy. As in the case of the relative entropy of Maxwell-Boltzmann statistics, the equality of the entropies before and after opening the window holds only approximately, because with the window open the particles might accumulate in one container, a possibility which can be neglected. For the proof of (38), this character of an approximation is again apparent from the fact that the last term in Stirling's formula (8, 6) has been neglected.

The mathematical aim of defining an absolute entropy satisfying the additivity postulate is therefore attained only in quantum physics.[3] Only in this physics, consequently, can Nernst's third law of thermodynamics be consistently interpreted as stating that the entropy is zero at the absolute zero point of temperature, because in quantum physics an absolute normalization of entropy can be carried through consistently.

Gibbs extended his paradox to the example of two nonidentical gases which are first separated and then merge by diffusion. In this case, the absolute entropy of Maxwell-Boltzmann statistics increases correctly to the value (8, 17). The same result would follow if the gases are similar; but for the limiting case of identical gases the entropy would be so normalized as to remain the same. This represents a discontinuity. In quantum physics, this discontinuity can be circumvented by the use of the concept of *partial likeness*; there will be a continuous transition of entropy values as the gases become more and more alike. This consideration has recently been used by A. Landé to derive the necessity of probability-controlled quantum jumps and other quantum-mechanical principles from a continuity postulate.[4]

• 28. Extension of Quantum Statistics to Different Energy Levels

We will now study the extension of the entropy concept to gas states, including various energy levels. For Maxwell-Boltzmann statistics

---

[3]This result has recently been emphasized by Alfred Landé in his book *Quantum Mechanics* (New York, Pitman Publ. Corp., 1951), pp. 219 and 227. The reader is also referred to this book for an excellent presentation of the physical side of quantum statistics and of quantum mechanics in general.

[4]See Alfred Landé, "Quantum Mechanics and Thermodynamic Continuity", *Amer. Jour. Phys.*, Vol. 20 (1952), pp. 353-358; his "Thermodynamic Continuity and Quantum Principles", *Phys. Rev.*, Vol. 87 (1952), pp. 267-271; and his recent book, *Foundations of Quantum Theory: A Study in Continuity and Symmetry* (New Haven, Conn., Yale Univ. Press, 1955).

this problem was discussed in §9. Let us briefly summarize this discussion. Representing the value of the energy associated with the state C_i by "ϵ_i", we add the energy condition to relations (27, 1-5), in the form

$$\sum_{i=1}^{n} \epsilon_i n_i = E \; . \tag{1}$$

The constant N is then to be replaced by the smaller but unknown constant N_t, as explained with respect to (9, 4). Writing the expression for the entropy S by the help of an undetermined constant, we thus replace (27, 5) by

$$S = n \log n - \sum_{i=1}^{m} n_i \log n_i + \text{const} \; . \tag{2}$$

We see that consideration of various energy levels merely changes the constant in S. Furthermore, S is now subject to the restrictive condition (1), which is added to (27, 3).

When we look for the maximum of S, the constant in (2) is irrelevant. It thus does not matter whether we use a relative or an absolute entropy. The computation carried out in §9 leads to the result (9, 11), which may be repeated here:

$$q(\epsilon) = \frac{n(\epsilon)}{m(\epsilon)} = ae^{-\epsilon/kT} \; . \tag{3}$$

This formula represents the average number of particles per state, for a given energy level, as depending exponentially on the temperature of the gas as a whole. In other words: out of the total number n of particles, a certain number $n(\epsilon)$ go into states of the energy level ϵ; at low temperatures, $n(\epsilon)$ is small for higher energy levels, but $n(\epsilon)$ increases for these higher energy levels with increasing temperature of the gas. The constant a, which can be evaluated only when the energy is given as a function of the parameters, decreases with growing temperature, as is shown for $b = a/n$ in (9, 15).

The state lattice, for which we have defined the quantity w_t, must now be constructed separately for every energy level; and formula (27, 28) applies when we use for q the value belonging to this energy level, given by (3).

In quantum statistics, in which entropy is defined with respect to a state lattice, and not, as in Maxwell-Boltzmann statistics, with respect to a particle lattice, we must begin the consideration by constructing a replica of the state lattice for every energy level. The number N_D of arrangements, given in (27, 6), is now to be applied to every energy level. Let us assume that the possible energy levels are given by $\epsilon^{(1)}, \epsilon^{(2)}, \ldots \epsilon^{(u)}$; all quantities associated with an energy level $\epsilon^{(s)}$ may be indicated by adding the corresponding superscript. Then the total number N_D of arrangements is given by the product

$$N_D = N_D^{(1)} \cdots N_D^{(u)} . \tag{4}$$

Correspondingly, the entropy (27, 11), or (27, 18), is replaced by a summation of values. Since here again the total number N of possible arrangements is to be replaced by the unknown number N_E, we will introduce into the measure of entropy an unknown constant.

Although these considerations apply both to Bose-Einstein statistics and to Fermi-Dirac statistics, we will consider for the present only the former. We write the general expression for the entropy in a gas of different energy levels, for Bose-Einstein statistics, as follows:

$$S = \sum_{s=1}^{u} m^{(s)} \log m^{(s)} - \sum_{s=1}^{u} \sum_{r=1}^{n^{(s)}} m_r^{(s)} \log m_r^{(s)} + \text{const} . \tag{5}$$

When we regard the additive constant as not depending on m, we have here an absolute-normalized entropy, which measures N_D, and not the probability W.

The restrictive conditions for S have now the form

$$\sum_{r=0}^{n} m_r^{(s)} = m^{(s)} , \tag{6} \qquad\qquad \sum_{s=1}^{u} \sum_{r=1}^{n} r\, m_r^{(s)} = n , \tag{7}$$

$$\sum_{s=1}^{u} \epsilon^{(s)} \sum_{r=1}^{n} r\, m_r^{(s)} = E . \tag{8}$$

Condition (6) stands for u similar conditions, each expressing the fact that summing up the $m_r^{(s)}$ of a certain energy level $\epsilon^{(s)}$ leads to the value $m^{(s)}$, which is a constant of the energy level. The other two conditions, (7) and (8), represent only one relation each, but have double summations. Note that the summation over r, in both relations, leads to the value $n^{(s)}$ of the energy level $\epsilon^{(s)}$; but since $n^{(s)}$ is not a constant of the energy level, it can only be required that the sum of the $n^{(s)}$ over s leads to the value n, which condition is formulated in (7).

The maximum of S in (5) is found, as in §9, by the use of a variation method employing Lagrange's multipliers. Since (6) represents u equations, it supplies u multipliers $\gamma^{(s)}$. Using α as multiplier for (7), and β for (8), we find, when we carry out the differentiation of (5)–(8) with respect to $m_r^{(s)}$:

$$1 + \log m_r^{(s)} + \gamma^{(s)} + \alpha r + \beta \epsilon^{(s)} r = 0 . \tag{9}$$

These $u \cdot n$ equations for the quantities $m_r^{(s)}$ can be written, when we go from logarithms to exponential forms:

$$m_r^{(s)} = g^{(s)} w^r ,$$

$$g^{(s)} = e^{-1-\gamma^{(s)}} , \qquad a = e^{-\alpha} , \qquad w = a \cdot e^{-\beta \epsilon^{(s)}} . \tag{10}$$

We write "w" instead of "$w^{(s)}$" as an abbreviation. In order to determine the constant $g^{(s)}$, we use condition (6), regarding n as infinite in the summation and applying (27, 34):

$$\sum_{r=0}^{\infty} m_r^{(s)} = m^{(s)} = g^{(s)} \sum_{r=0}^{\infty} w^r = g^{(s)} \cdot \frac{1}{1-w} ,$$

$$g^{(s)} = m^{(s)} \left(1 - w\right) . \tag{11}$$

Using the considerations applied for (9, 8–10), we find that $\beta = 1/kT$. Furthermore, we will now compute the quantity $n^{(s)}$ mentioned before, which expresses the number of particles on the energy level $\epsilon^{(s)}$:

$$n^{(s)} = \sum_{r=1}^{\infty} r\, m_r^{(s)} = m^{(s)} \left(1-w\right) \sum_{r=1}^{\infty} r\, w^r = m^{(s)} \cdot \frac{w}{1-w} . \tag{12}$$

In (12) we have used formulas (10) and (11) and a formula for geometrical series; this third formula, which is easily derived by differentiating formula (27, 34) with respect to w, is as follows:

$$\sum_{r=0}^{\infty} r\, w^r = \sum_{r=1}^{\infty} r\, w^r = \frac{w}{(1-w)^2} , \qquad 0 \leq w < 1 . \tag{13}$$

When we now introduce (11) into (10) and solve (12) for w, we find, defining $w_r^{(s)}$ by analogy with (27, 19):

$$w_r^{(s)} = \frac{m_r^{(s)}}{m^{(s)}} = \left(1-w\right)w^r , \qquad w = \frac{n^{(s)}}{n^{(s)} + m^{(s)}} . \tag{14}$$

We thus find that formula (27, 33) holds for every energy level $\epsilon^{(s)}$ separately. This formula, which we derived, in the form (27, 29), from mere combinatorics, is now valid only in the sense of a maximum probability, that is, for stationary conditions. The reason is that only in the stationary state will $n^{(s)}$ have the value given in (12).

However, the given derivation has presented us with an important addition to the formulas (27, 29) and (27, 33): it gives us the value of w, or q, as a function of the energy level ϵ. To write this relation, we combine the value of w from (10) with the one from (14), now writing "$w(\epsilon)$" and replacing the superscript "(s)" throughout by the argument "ϵ" added to the quantities. Using the relation (27, 32) between w and q, we thus find:

$$\frac{n(\epsilon)}{n(\epsilon) + m(\epsilon)} = \frac{q(\epsilon)}{1 + q(\epsilon)} = w(\epsilon) = ae^{-\epsilon/kT}. \tag{15}$$

This relation formulates, for Bose-Einstein statistics, the dependence of the particle number $n(\epsilon)$, or of the related numbers $q(\epsilon)$ and $w(\epsilon)$, on the energy level. It corresponds to the relation (3), of Maxwell-Boltzmann statistics. Like the latter relation, it states an exponential dependence, though of a slightly different form. However, if $q(\epsilon)$ is small, we have $q(\epsilon) \sim w(\epsilon)$, and relation (15) is then virtually identical with (3). The constant a, too, is then the same for both relations, because this constant is determined by summing the exponential expression over all energy levels. Note that the constant a contains the factor n. In addition, a depends on the temperature T, as is illustrated by the special form (9, 15) for $b = a/n$.

Corresponding derivations can be made for Fermi-Dirac statistics. Using (27, 13–18) and summing over the various energy levels, we find

$$S = \sum_{s=1}^{u} m^{(s)} \log m^{(s)} - \sum_{s=1}^{u} m_0^{(s)} \log m_0^{(s)} - \sum_{s=1}^{u} m_1^{(s)} \log m_1^{(s)} + \text{const} , \tag{16}$$

$$(17) \qquad m_0^{(s)} + m_1^{(s)} = m^{(s)} , \qquad\qquad m_1^{(s)} = n^{(s)} , \tag{18}$$

$$(19) \qquad \sum_{s=1}^{u} m_1^{(s)} = n , \qquad\qquad \sum_{s=1}^{u} \epsilon^{(s)} m_1^{(s)} = E . \tag{20}$$

When we maximize S, we find relations (9) and (10); however, they represent here only $2u$ relations. Instead of (11), we find

$$m_0^{(s)} + m_1^{(s)} = m^{(s)} = g^{(s)} (1 + w) , \tag{21}$$

$$g^{(s)} = \frac{m^{(s)}}{1 + w} .$$

Instead of (12), we have here

$$n^{(s)} = m_1^{(s)} = m^{(s)} \frac{w}{1 + w} \; . \tag{22}$$

Inserting these values into (10) and putting there $r = 0$ and $r = 1$, respectively, we find, using (18) and (27, 31):

$$w_0^{(s)} = \frac{m_0^{(s)}}{m^{(s)}} = \frac{1}{1 + w} = 1 - q^{(s)} \; , \tag{23}$$

$$w_1^{(s)} = \frac{m_1^{(s)}}{m^{(s)}} = \frac{n^{(s)}}{m^{(s)}} = \frac{w}{1 + w} = q^{(s)} \; . \tag{24}$$

We thus find that relations (27, 30) apply to every energy level separately, in analogy to (14). When we now use the argument "ϵ" instead of the superscript "(s)" and introduce the value of w from (10), we find, in analogy to (15):

$$\frac{n(\epsilon)}{m(\epsilon) - n(\epsilon)} = \frac{q(\epsilon)}{1 - q(\epsilon)} = w(\epsilon) = a e^{-\epsilon/kT} \; . \tag{25}$$

This relation formulates the exponential dependence of $n(\epsilon)$ on the energy level for Fermi-Dirac statistics. The difference from (15) consists only in the sign occurring in the denominators of the first and second expression. This makes a great difference for large values $q(\epsilon)$; but for small values $q(\epsilon)$, formulas for the two kinds of quantum statistics (Bose-Einstein and Fermi-Dirac) converge to the exponential formula (3) of Maxwell-Boltzmann statistics.

By solving equations (15) and (25) for q, these relations are often combined in the following form:

$$q(\epsilon) = \frac{n(\epsilon)}{m(\epsilon)} = \frac{1}{a^{-1} e^{\epsilon/kT} \mp 1} \; . \tag{26}$$

The minus sign applies to Bose-Einstein statistics, the plus sign to Fermi-Dirac statistics. If the constant a is small, its reciprocal value a^{-1} is large; then the term ∓ 1 can be neglected, and formula (26) becomes identical with the Maxwell-Boltzmann form (3).

We can now answer the question, raised at the beginning of §27, concerning the gradual transition from quantum statistics to Maxwell-Boltzmann statistics. Equations (3), (15), and (25) tell us that for small values $q(\epsilon)$ the three kinds of statistics furnish virtually the same number $n(\epsilon)$ of particles per energy level. Now the discussion of relations (27, 28–30) showed, as was seen from the discussion of

(27, 35) and (27, 36), that for equal $n(\epsilon)$ and small $q(\epsilon)$ the probabilities w_r are essentially the same for the three kinds of statistics. Thus small $q(\epsilon)$ lead to equality of the three types of statistics, both with respect to number of particles per energy level and with respect to distribution of particles within an energy level. Therefore quantum statistics and Maxwell-Boltzmann statistics converge in every respect, when $q(\epsilon)$ goes to low values; this includes a convergence of the entropy values.

For low values q, the occupation numbers are virtually all 0 or 1, and we have $m_r \sim 0$ for $r > 1$. This applies both to the two kinds of quantum statistics and to Maxwell-Boltzmann statistics; thus the distribution of maximum probability is the same for all three kinds of statistics. A difference results merely for the numerical value of the entropy; using Maxwell-Boltzmann statistics, we would interpret each arrangement of the maximal distribution as consisting of N_D Maxwell-Boltzmann arrangements, where N_D is given by (27, 37). The latter value reduces to $n!$, if $m_r = 0$ for $r > 1$. The total number of arrangements would therefore result from multiplying the quantum-statistical number by $n!$. As a consequence, the entropy values (27, 11) of Bose-Einstein statistics and (27, 18) of Fermi-Dirac statistics differ from the entropy (27, 5) of Maxwell-Boltzmann statistics only by the additive constant $\log(n!)$. The same applies to the generalized entropy values (5) and (16) in comparison with (2). Here the additive constant has the form "$\sum_{s=1}^{u} \log [n^{(s)}!]$". Since the $n^{(s)}$ are practically the same for all three kinds of statistics if q is small, this constant has the same value for them; and their entropies are again identical, except for an additive constant. For small values q, the definition of entropy no longer discriminates between the three kinds of statistics.

Now $q(\epsilon)$ is small when the constant a in equations (3), (15), and (25) is small. Since a contains n as a factor, the total number n of particles must be small as compared with the number m of accessible states. Furthermore, the value of a decreases when the temperature increases; this inverse relation, however, depends on the nature of the gas atoms, that is, on the form of the energy ϵ as a function of the parameters. Since increase in temperature, furthermore, means increase in energy and thus increase in the number of accessible states, we can conclude that for higher temperatures the three kinds of statistics furnish practically the same observational results.

For the usual gas, room temperature is a high temperature; but for temperatures close to the absolute zero point, peculiar deviations in gas behavior are observed which translate the differences between the three kinds of statistics into observational results. The degeneration of liquid helium, which follows Bose-Einstein statistics, was mentioned in §26. For the "gas" of conduction electrons in a wire, the Fermi-Dirac behavior of which was mentioned in §26, room temperature is a low temperature. Thus observational techniques allow us to show that when observable differences can be expected quantum statistics is confirmed, whereas Maxwell-Boltzmann statistics can be applied only under those normal conditions for which all three kinds of statistics lead to the same observational results. The applicability of Maxwell-Boltzmann statistics must therefore be understood as representing an approximation which for normal physical conditions cannot be distinguished from the more precise form of Bose-Einstein statistics.

This consideration explains why the same particles can be regarded as distinguishable for higher temperatures, whereas they must be regarded as indistinguishable for lower temperatures. A transition to higher temperatures does not equip particles with specific marks, so as to make them individually distinguishable. We should rather say that the particles remain indistinguishable even though the temperature rises, but that their statistical effects are eventually the same as though they were distinguishable: it then is no longer possible to distinguish between indistinguishability and distinguishability. This conclusion reveals a characteristic asymmetry between the two conceptions: whereas statistical results at low temperatures compel us to say that particles are indistinguishable, the statistical results at higher temperatures do not prove that particles are distinguishable, but merely allow us to treat particles as distinguishable.

The question may be raised whether under normal temperature and pressure conditions the number q of particles per state is as small as is needed to make the three kinds of statistics indistinguishable from one another. In classical kinetic theory, it is usually assumed that the volume of a gas is divided into small cells, which cells are, however, each large enough to contain a large number of molecules. If these cells were regarded as the compartments of our statistics, Bose statistics would lead to results noticeably different from those for Boltzmann statistics. But such cells are too large to represent states in the quantum-physical sense. They are aggregates of quantum states and must be treated as such. When we come to cells of quantum size—for instance, when we regard the kinetic energy as quantized—the number of particles per cell is as small as is necessary for a convergence of quantum statistics to classical statistics.

The statistics of an aggregate of quantum states differs greatly from that of individual states. When we divide a room of a house into cubicles of one cubic millimeter each, we find virtually the same number of air molecules in every cubicle. In classical statistics the inference was made that the same homogeneity would be observed if we divided the room into still smaller cells. For very small cells, fluctuations were assumed, which, however, were supposed to manifest themselves in small deviations from the average occupation number n/m. Using Bose statistics, we would explain the observational result very differently. We would then regard our cubic-millimeter cells as composed of an enormous number m' of quantum cells; for these very small cells, we have not a homogeneous distribution, but the inhomogeneous distribution given by the w_r of Bose statistics, for which the molecules are crowded together in certain cells and the fluctuations are much larger than for Maxwell-Boltzmann statistics. Since a cubic-millimeter cell is composed of so very many quantum cells, we have for the number n_i' of molecules in it, using the $w_r^{(s)}$ of (14) for w_r and applying (13),

$$ n_i' = m' \cdot \sum_{r=1}^{\infty} r \, w_r = m' \cdot \frac{w}{1-w} = m' \cdot q \; . \tag{27} $$

For the various cubic-millimeter cells, the numbers n_i' are virtually the same, because of the law of large numbers, according to which fluctuations in n_i' decrease with the square root of m'. The homogeneous distribution in large cells thus appears to be the result of statistical compensation of inhomogeneities in much smaller cells, and it seems that this distribution would still be observed even though the number q of particles per quantum state were not small. Thus, aggregates of quantum states behave like Maxwell-Boltzmann states.

These considerations may be illustrated by a numerical example. Suppose we have $m = 10$ compartments and $n = 100$ particles; then we we have $q = n/m = 10$. When we apply Bose-Einstein statistics, the probability that a compartment is empty is $w_0 = 1/(1+q) = 1/11 \sim 1/10$. If the distribution process is often repeated, this means that on the average about one compartment is empty. Now suppose that each compartment is subdivided into 100 small cells. Relative to these 1,000 cells, we have $q = 1/10$, and Bose statistics supplies $w_0 = 10/11$ for the probability that one of the small cells is empty. In order that one of the larger compartments be empty, we must have 100 empty cells in a row; therefore the probability that this occurs is now equal to $(10/11)^{100}$, which is of the order of 10^{-5} and negligible. This means we now have practically no empty compartments.

Applying Maxwell-Boltzmann statistics, for the first situation in which the compartments are not subdivided, and using the strict form of formula (27, 28) (given in note 2 of §27 as formula [28a]), we find $w_0 = [(m-1)/m]^n = (9/10)^{100}$, which is practically identical with the last value computed for Bose-Einstein statistics. When we subdivide the compartments, we find $w_0 = (999/1,000)^{100}$ for the probability that a cell is empty; this is practically 9/10, when we approximate $(1-\delta)^n$ by $1-n\delta$. The probability that a compartment is empty is again equal to the probability that 100 cells in a row are empty; this probability is $(9/10)^{100}$ and thus the same as before.

For Maxwell-Boltzmann statistics, the size of the compartments, or size of states, is therefore irrelevant; in contrast, Bose-Einstein statistics depends on the absolute size of the states. In precise language, we must speak of *Bose statistics with respect to a certain set of compartments.* This conclusion offers no difficulties, since in quantum physics the states, in fact, are given in an absolute size (see §26). An aggregate of states has other properties than a single state has, because its statistics results from a combination of counting indistinguishable particles and distinguishable states. If the aggregate contains a large number of states, its properties are essentially determined by the statistics of distinguishable states and thus resemble those of Maxwell-Boltzmann statistics.

These considerations explain, furthermore, why quantum statistics becomes identical with Maxwell-Boltzmann statistics for the limiting case $h = 0$. If the quantum cells had the size 0, Bose statistics with respect to these cells would be identical with Maxwell-Boltzmann statistics for any cells of nonvanishing size. More precisely speaking, this limit statement has the following meaning: given a set of compartments C_i, then it is possible to find a set of smaller compartments $c_i = (1/k)C_i$ such that Bose statistics with respect to the set of c_i is equivalent, within a given degree δ of exactness, to Maxwell-Boltzmann statistics with respect to the set of C_i.

We see that it represents an invalid extrapolation to infer from the homogeneous distribution in large cells that the same homogeneity applies to smaller cells, and that the smaller cells, too, satisfy the Maxwell-Boltzmann distribution. Homogeneity on a larger scale is compatible with inhomogeneity on a smaller scale; therefore, macroscopic homogeneity does not allow us to discriminate between distinguishable and indistinguishable particles. Since the observations of quantum phenomena, which refer to experiments in which the pecularities of quantum statistics are projected into the macrocosm, decide in favor of the second alternative, we must regard the apparent homogeneity of the macrocosm as resulting from statistical compensation rather than from similar homogeneities in the microcosm.

• *29. Particles Vanishing into Nonexistence*

The mathematical results derived in §§27 and 28 may be used to answer a question which is related to the logical analysis of genidentity given in §26. It was pointed out in this analysis that, precisely speaking, we do not have a proof that quantum particles are indistinguishable; we can only prove that they must be so interpreted if we are to avoid causal anomalies. The question can be raised whether there exists for Bose and Fermi statistics an interpretation, in terms of distinguishable particles, for which the causal anomalies do not have the form of mutual dependence relations though they may assume different forms. The present section is intended to show that the answer is affirmative, provided the genidentity of particles is construed in a peculiar way.

This interpretation may first be developed for Fermi statistics. We will introduce the assumption that a particle leads two lives; sometimes it is "active" and noticeable, sometimes it is "frozen" and cannot be noticed through physical interaction with other things. Let us further assume that one and the same particle stays in one state all the time, however, changing its condition at random; sometimes it is in a frozen condition, sometimes in an active condition. If the particle is frozen, it vanishes temporarily into nonexistence, so to speak; and the state is in the condition which normally would be called "empty". The probability that a particle jumps into activity is equal to q; the probability that it jumps into a frozen condition is equal to $1 - q$. These probabilities correspond to formula (27, 30); and in accordance with (28, 25) we assume that q, which is a constant for every state, is determined exponentially by the energy ϵ. The world line of a particle, divided by small intervals Δt, is then a random probability sequence governed by the probability q. The particles are all independent of one another and no exclusion principle is required. Instead, we now have the anomaly that a particle is subject to two different conditions, one of which is a frozen condition.

A similar interpretation can be carried out for Bose statistics. The expression (27, 33) for w_r suggests here an interpretation in terms of runs of a random sequence. We will study this interpretation first on the assumption that all states belong to the same energy level, for which case (27, 33) was developed.

When we play for an event B with the probability w in a random sequence, and make a dividing line after every "$\bar{B}$", we arrive at a pattern of the following kind:

$$B\ B\ \bar{B}|B\ B\ B\ \bar{B}|\bar{B}|B\ \bar{B}|B\ \bar{B}|B\ B\ \bar{B}|\ldots \tag{1}$$

Each section contains a run of exactly r elements "B" and ends with a "$\bar{B}$". The probability of such a run is given as follows:

$$P\left(\bar{B}, B^1 \ldots B^r . \bar{B}^{r+1}\right) = \left(1-w\right)w^r = w_r \ . \tag{2}$$

The probability of a run of r elements B is counted in the series of sections; this way of counting is assured through the term "$\bar{B}$" in the reference class of (2), because every "$\bar{B}$" introduces a subsequent section. If a section contains only the element "$\bar{B}$", we have $r = 0$. The value w_r of (2) is the value (27, 33) of Bose statistics.

This result allows us to construct the following macrocosmic model of Bose statistics. We have a bag filled with balls and a number of compartments, or states. We employ some mechanism supplying the probability w for an event B, and start playing. If B turns up, we put a ball from the bag into the first compartment, and go on playing. If, on playing the second time, B turns up again, we put another ball into the same compartment. Thus we go on until a $\bar{B}$ turns up. Then we put no ball into the compartment, and we close the filling process for this compartment. We go on playing with the probability w, now filling the second compartment by the same method. When a $\bar{B}$ follows a $\bar{B}$, the compartment gets no ball at all. In this way we continue until we have accounted for all the compartments; the resulting occupation numbers then follow Bose statistics.

There exists a second form for the filling method. When $\bar{B}$ turns up, we still put a ball into the compartment, but mark the ball as "frozen". Then the filling of this compartment is closed and we go on playing for the next compartment. In this way, every compartment receives just one "frozen" ball, in addition to the "active" balls which supply Bose statistics. That is, we now have occupation numbers $r' = r + 1$ such that $w'_{r'} = w_r$.

We will use the second form to construct a particle lattice through repetition of the procedure, still holding to the assumption that all states belong to the same energy level. Suppose we have m compartments and n balls; if these numbers are large, we may assume that in every repetition all n balls are used and all m compartments are filled. We will furthermore write individual numbers on all the balls, in order to identify them. But the balls are always drawn at random from the bag. The compartments may be selected in some definite order, which is the same all the time, or also in a random order, which varies with every filling process. When the first filling procedure is finished, we write down for every ball into which compartment it went, and whether or not it is there in an active condition B. Then we take all the balls out, erase the "frozen" marks, but leave the individual numbers on

TABLE 7

PARTICLE LATTICE FOR BOSE STATISTICS

(This lattice gives the attributes "active" $[B]$ and "frozen" $[\bar{B}]$, and in addition the state C_i into which the particle went. The event B is controlled by the probability w, both in rows and in columns.)

	t_1	t_2	t_3	t_4	$\cdots$
Particle 1	B, C_{12}	B, C_1	$\bar{B}, C_4$	B, C_4	$\cdots$
Particle 2	B, C_2	$\bar{B}, C_1$	B, C_{21}	B, C_2	$\cdots$
Particle 3	$\bar{B}, C_{12}$	$\bar{B}, C_4$	B, C_{67}	B, C_{14}	$\cdots$
Particle 4	B, C_{12}	B, C_1	$\bar{B}, C_{12}$	B, C_4	$\cdots$
$\cdots\cdots\cdots$	$\cdots\cdots$	$\cdots\cdots$	$\cdots\cdots$	$\cdots\cdots$	$\cdots\cdots\cdots$
Particle n	B, C_{20}	B, C_4	B, C_{67}	B, C_1	$\cdots$

the balls, and begin anew. Again the balls are selected in a random order. Going on in this way, we arrive at a lattice of the form given in table 7.

In the lattice thus resulting, every row represents the history of a particle. With respect to the attributes B and $\bar{B}$, all rows and columns are random sequences controlled by the probability w and independent of one another; we thus have here a lattice of normal sequences in the narrower sense.[1] In contrast, the attributes C_i supply Bose statistics. This means: when we count in a column the number $n_i = r$ of particles showing the same attribute C_i, the frequency of these numbers r, which include one frozen particle for each compartment, is controlled by the value w_{r-1} of (2). More precisely speaking: if m_r is the number of compartments in a column all having the occupation number r, we find $w_{r-1} = m_r/m$. This holds within every column.

This lattice permits of the following physical interpretation. The life history of a particle includes frozen and active conditions. On each step, the particle jumps anew: the probability that it jumps into an active condition is equal to w, and the probability that it jumps into a frozen condition is equal to $1 - w$. In these jumps all the particles are independent of one another. Assume that the states are filled at a certain moment. In the next moment, all particles, including the frozen ones, leave their states and jump for new states. When a particle jumps into a new state, it has the probability w to be there in an active condition; this applies, too, if the particle was frozen in the previous state. But once a particle jumps into a state and enters into a frozen condition, this state is closed for all further particles. Such a redistribution occurs from moment to moment.

Through this interpretation, we have constructed a particle lattice

---

[1]*ThP*, p. 172.

in which there are no causal anomalies that express a causal depend-ence. The world lines of the particles are independent as far as the attribute B is concerned, and the Bose statistics with respect to the attribute C_i results merely from the process in which a frozen particle seals off a state from further particles. This exclusion condition, however, represents a specific causal anomaly. And though we have here particles of physical individuality, their genidentity is anomalous, inasmuch as each life history includes periods during which the par-ticle is unobservable, in which it "exists in the form of nothingness".

An extension of these conceptions to states of different energy levels can be constructed as follows. The probability w that a particle is active depends now on the energy level associated with the state, according to (28, 15). In the filling procedure of the macrocosmic model, we use for every compartment a corresponding value for w. The rows of the lattice of table 7 are then sequences of varying prob-abilities; a specific constant value for w appears only in the subse-quences selected by compartments C_i of the same energy level. The same kind of variation of w applies to the columns.[2]

The physical interpretation is the same as before. It should be noted that this interpretation does not constitute a physical hypothesis and is not presented here as a proposal to physicists to change their con-ception of Bose statistics. The interpretation given represents merely a mode of speech. The assumptions on which it is based (the jumping of the particles into frozen or active conditions, the sealing off of each state by the particle that freezes in it, and the exponential de-pendence of the probability that a particle is active on the energy level) are all true by definition. They represent conventions on which this interpretation is based; these conventions are admissible, because they are compatible with the observed statistics. The interpretation shows that when we wish to introduce material genidentity into gas statistics, the resulting causal anomalies are still subject to our choice.

• 30. Particles Traveling Backward in Time

The generalization of the genidentity concept, which led to a mere functional genidentity of indistinguishable particles, has still left a modicum of individuality to the elementary particles, since the number of particles remains constant during the transition to other distribu-tions. If particles cannot be individually distinguished, they can at

---

[2]Sequences in which the probability varies from element to element have often been studied; see *ThP*, p. 168.

least be counted. It turns out, however, that even this remnant of individuality can be questioned.

The conservation of number has been abandoned in the theory of photons, that is, the theory of radiation. The energy of a photon of the wave length λ, or the frequency ν, is equal to $h\nu$, according to M. Planck's fundamental relation. If a closed container is filled with radiation, which is absorbed and reëmitted by the black walls of the container, an exchange of wave lengths takes place; since the photons resulting from this exchange carry an energy different from that of the original photons, the number of photons cannot be constant if the sum total of their energies is to remain constant. In fact, when we apply to photons the maximization method described in §28 and omit condition (28, 7), which states the conservation of number, the constant α in (28, 9) equals 0, and, consequently, we find from (28, 10) that $a = 1 = a^{-1}$. Introducing this value in (28, 26) for Bose statistics and replacing "ϵ" by "$h\nu$", we arrive at the formula

$$q\left(\epsilon\right) = \frac{n(\epsilon)}{m(\epsilon)} = \frac{1}{e^{-h\nu/kT} - 1} \quad . \tag{1}$$

This is Planck's famous radiation law, with which the quantum theory made its entrance into physics. Planck himself did not derive his law in this way, but by means of classical statistics of linear oscillators. But Bose statistics, devised a quarter of a century later, yields the same result if radiation is regarded as consisting of indistinguishable particles whose number is not preserved.

Now photons are not material particles in the ordinary sense. They have no rest mass; that is to say, if they could be stopped in their motion, they would have no mass at all. They owe all their mass to moving with the speed of light. Therefore it is not too surprising that their number is subject to change. It would be different if such material particles as electrons could be shown to vary in number within a closed container.

Surprisingly enough, recent developments have demonstrated that the genidentity of material particles can be questioned more seriously than is done in Bose statistics. The difference between one and two, or even three, material particles can be shown to be a matter of interpretation; that is, this difference is not an objective fact, but depends on the language used for the description. The number of material particles, therefore, is contingent upon the extension rules of language. However, the interpretations thus admitted for the language of physics differ in one essential point from all others: they require an abandonment of the order of time..

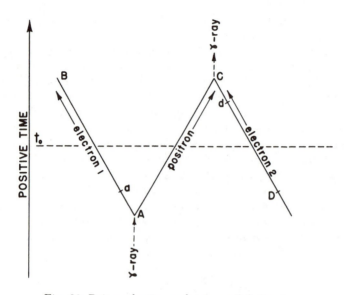

Fig. 38. Pair production and pair annihilation in a
Wilson cloud chamber.

Conceptions of this kind were developed by E. C. G. Stückelberg and
R. P. Feynman.[1] Their investigations showed that a positron—that is,
a particle of the mass of an electron, but carrying a positive unit
charge—can be regarded as an electron moving backward in time. The
negative unit charge of the electron, which travels in the opposite
time direction, has the same physical effects as the charge of the
positron traveling forward in time; and therefore the two interpretations
cannot be distinguished observationally.

Feynman showed that these conceptions can be used for an explanation of pair production and pair annihilation. It has been observed on
photographs taken in a Wilson cloud chamber that, upon incidence of
a γ-ray, an electron and a positron are generated from "nothing" and,
starting from the same point, travel along different paths. The positron
is usually not long-lived; it encounters some other electron traveling
free through space and then merges with it in an act of collision.
These two particles thus vanish completely, leaving as their effect

[1]See E. C. G. Stückelberg, "Remarque à propos de la création de paires de
particules en théorie de relativité", *Helv. phys. Acta*, Vol. 14 (1941), pp.
588–594; and "La mécanique du point matériel en théorie de relativité et en
théorie des quanta", *ibid.*, Vol. 15 (1942), pp. 23–37. Also see R. P. Feynman,
"The Theory of Positrons", *Phys. Rev.*, Vol. 76 (1949), pp. 749-759.

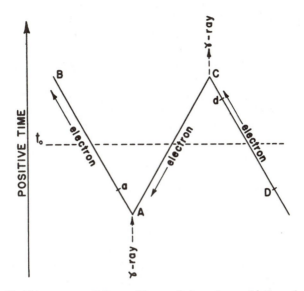

Fig. 39. The process of figure 38 regarded as the world line of a single electron, which from *C* to *A* travels backward in time.

only a new γ-ray starting from the point of collision. Figure 38 may illustrate these processes. Positive time is represented by a vertical line going upward; the other solid lines represent world lines of particles. Dotted lines indicate the world lines of the γ-rays. In the event *A*, the incident γ-ray produces a pair consisting of electron number 1 and a positron. In the event *C*, the positron collides with electron number 2; this pair is annihilated in the collision, the only trace being the γ-ray starting at *C*. In the photograph, the paths of the particles are visible and show a spatial arrangement similar to that of the solid lines in the diagram; the γ-rays are not visible in the photograph.

According to Feynman, we can as well interpret the process diagramed in figure 38 by regarding the train of lines *DCAB* as the world line of one single electron, which from *C* to *A* travels backward in time, as indicated in figure 39. Instead of three particles, we thus have only one. This interpretation has the advantage that we need not speak of pair production and pair annihilation; the one particle is there all the time. The causal anomalies of creation from nothing and vanishing into nothing are thus eliminated; however, in exchange for them another causal anomaly enters the description: the electron travels part of its path backward in time.

We meet here with a new illustration of the theory of equivalent descriptions in application to interphenomena, that is, to unobservables. The photograph shows us the picture of small droplets of water in a row; we interpret them as produced by a particle which hits larger particles along its path and induces condensation of water vapor. The traveling particle is inferred, and thus represents an interphenomenon. In the usual description, we say that the interphenomenon consists in a positive particle traveling forward in time. In the new description, we say that it consists in a negative particle traveling backward in time. The succession of impacts manifest in droplets thus keeps its time sequence, macroscopically speaking; but seen from the particle, so to speak, these impacts occur in the opposite sequence. This mode of speech, referring to interphenomena only, permits us to eliminate the anomalies of pair production and pair annihilation, because, seen from the particle, there are no pairs, there is only one particle. But the elimination of the anomaly is dearly paid for by the emergence of another anomaly, the travel backward in time. As always in quantum mechanics, an exhaustive description of interphenomena is associated with causal anomalies; we have merely the choice where to place them. We may recall here the results arrived at in §29, where we found that it is sometimes convenient to assume that particles vanish temporarily into nonexistence. The present example deviates from a normal causal interpretation of interphenomena in another way: it uses a physical description in terms of a reversal of causal world lines.

Although there is no pair production or pair annihilation for the new description, there still exist the two γ-rays, which are observationally verifiable and thus cannot be "interpreted away". These rays are here regarded as causally connected with the reversal of time at the two point events A and C; reversal is here defined as deviation from the time direction of the environment. The reversal process can be accounted for mathematically by a suitable adjustment of equations. Considerations of this kind were first carried out by Stückelberg and became the starting point of the new interpretation.

It should be noted that this interpretation does not merely signify a reversal of time *direction*; it represents an abandonment of time *order*. In our previous discussion of time direction, we have occasionally envisaged a reversal of all causal chains and studied the picture which the world would then offer by imagining a film reeled in reverse. In identifying a positron with an electron going backward in time, however, only one causal chain is reversed, while the others keep their direction. As a consequence, the between-relations (see §22) are changed. In figure 39, point event C is causally between D and A; and point event A is causally between C and B. In figure 38, this is not so. It is this change of between-relations which we refer to when we speak of abandonment of time order.

In giving up time order, we open the way for the possibility of closed causal chains. Applying the language of the usual interpretation (see fig. 38), suppose that at a, electron number 1 enters into some collision process accompanied by the emission of a light ray, which travels faster than the positron and hits electron number 2 at d. This electron then need not have come from D, but we will assume that the segment dC of its path is the same as before. When we now introduce the second interpretation (fig. 39), the light ray ad will not be reversed; and the train $dCAad$ will now constitute a closed causal chain. Such processes have not been observed; but though they appear rather improbable, their possibility cannot be denied.

Closed causal lines of this kind do not offer the difficulties arising when macroscopic causal lines are closed. We saw in §18 that the concept of producing is a statistical concept; this concept, therefore, cannot be applied to atomic chains, and we cannot say that here the effect produces its own cause. Likewise, the mark concept is statistical (see §23); therefore the closed atomic chain cannot lead to paradoxes concerning the transfer of a mark. And we cannot apply the concept of recording, either, because it, too, is of a statistical nature. Regarding A as the effect of C, as is done in the second interpretation, we cannot tell any more about C from observing A than we can predict about C in the usual interpretation. A closed causal line of the atomic kind does not disturb our concept of causality, which is a macrocosmic concept; but it does contradict the concept of time order, and with it, the usual concept of genidentity.

In fact, with the identification of the positron with an electron traveling in negative time, the concept of physical identity is shaken to its very foundation. One and the same individual can exist in more than one specimen at the same time; the relation of genidentity is no longer unique among simultaneous events. Bose statistics had required us to give up material genidentity; but the functional genidentity put into its place still satisfied the rule that among simultaneous states no two states can belong to one particle. It is different for the genidentity of positrons and electrons. If we draw a simultaneity cross section in terms of macrocosmic time, as indicated by the horizontal dotted line t_0 in figures 38 and 39, then at this time the electron of Feynman's interpretation exists in three specimens, given by the three states s_1, s_2, and s_3, whereas in the usual interpretation the three states s_1, s_2, and s_3 belong to three different particles, one being a positron. Yet seen from the causal line $DCAB$ of figure 39, the path going from C to A is not backward in time. In terms of the "proper time" of the electron, this, too, like the other parts of the train, is a forward path; and the three states s_3, s_2, and s_1 are not simultaneous, but follow one another in this order. However, what is temporally forward for the electron is temporally backward for other things—in particular, for the macroscopic apparatus given by the mechanism of the Wilson cloud chamber. We have here individual definitions of time direction which cannot be pieced together into a consistent net; for this reason, there exists no time order for such occurrences.

This is the most serious blow the concept of time has ever received in physics. Classical mechanics cannot account for the direction of time; but it can at least define a temporal order. Statistical mechanics can define a temporal direction in terms of probabilities; but this definition presupposes time order for those atomic occurrences the statistical behavior of which supplies time direction. Quantum physics, it appears, cannot even speak of a unique time order of its processes, if further investigations confirm Feynman's interpretation, which is at present still under discussion. What all this means for quantum physics is unknown today. Perhaps it means that the variable "t", which has often played a questionable part in quantum physics, has only a "local" meaning and cannot always be extended to larger domains; perhaps the local t must be distinguished from the general t which is the time of macrocosmic processes. We shall have to wait for a consistent theory of the atomic nucleus in order to find the answers to these questions.

One conclusion, however, can already be drawn. Even time order, not only time direction, of the macrocosm must be accounted for by statistical methods. A way of doing this, in principle, was indicated in §22. The results of quantum physics offer us another possibility of explaining time order. Positrons are short-lived and therefore constitute only a small fraction of all the particles existing at a macrocosmic moment. The vast majority of particles thus conform to the rules of an ordered time; and time order springs from the atomic jumble by statistical dominance of negative electric particles. If it were different, if there were as many positrons as there are electrons, it would be doubtful whether an ordered time would exist; maybe there would exist closed causal lines in the macrocosm. We have to conclude that the existence of a linear time order is connected with the asymmetry of negative and positive electricity. According to conceptions developed by Dirac, this asymmetry originates in the fact that practically all states of negative energy are occupied, whereas those of positive energy are in general empty; the positron may be imagined as a sort of hole left by an electron which has abandoned its place on a negative energy level. This is picture language, but it expresses ideas which can be formulated in mathematical equations. A logical analysis of these problems is highly desirable. But the supposition appears plausible that we owe the existence of an ordered macrocosmic time to the peculiar difference between the two forms of electricity.

The conclusion of this brief survey of recent developments is that quantum physics does not present us with the time direction denied to us by the atomic occurrences of classical physics. Quantum physics worsens the situation; it makes even time order, not only time direction, a statistical property. Time appears to be a completely macrocosmic phenomenon, which cannot be traced into the microcosm; it is born anew at every moment from the atomic chaos as a statistical relationship. Strangely enough, this origin from disorder does not make macrocosmic time inferior. On the contrary, it will be seen that the birth from an atomic chaos endows the statistical cosmos with a time of exactly those properties which common sense and everyday experience have always regarded as intrinsic characteristics of temporal flow.

• *Appendix*

[The plan of the book contained a final chapter which would have dealt with the relation between the subjective experience of time by human beings and the objective properties of time in nature. In order to indicate some of these ideas I give here a translation of a few paragraphs from Hans Reichenbach, "Les Fondements logiques de la mécanique des quanta", *Annales de l'Institut Henri Poincaré*, Tome XIII, Fasc. II (Paris, 1953), pp. 156-157. —M. R.]

The distinction between the indeterminism of the future and the determinism of the past is expressed in the last analysis in the laws of physics. This is the important result of the combination of classical statistics with the indeterminacy relation of quantum physics. The consequences for the time of our experience, that is, the time of everyday life, are obvious. The concept of *becoming* acquires a meaning in physics: The present, which separates the future from the past, is the moment when that which was undetermined becomes determined, and "becoming" means the same as "becoming determined".

One question remains to be discussed. What is the relation between the time of physics and the time of our experience? Why is the flow of psychological time identical with the direction of increasing entropy?

The answer is simple: Man is a part of nature, and his memory is a registering instrument subject to the laws of information theory. The increase of information defines the direction of subjective time. Yesterday's experiences are registered in our memory, those of tomorrow are not, and they cannot be registered before the *tomorrow* has become *today*. The time of our experience is the time which manifests itself through a registering instrument. It is not a human prerogative to define a flow of time; every registering instrument does the same. What we call the time direction, the direction of becoming, is a relation between a registering instrument and its environment; and the statistical isotropy of the universe guarantees that this relation is the same for all such instruments, including human memory.

Let us add a remark concerning the term "now". Symbolic logic tells us that we must distinguish between the individual sign [token] and the class of signs, the symbol. With regard to many words one can neglect the individual sign; the word "house", for instance, has the same meaning in all instances. It is different for words like "now", "here", "I", the meaning of which changes with every individual sign. I should like to speak in this case of token-reflexive signs and token-reflexive meaning, using the word "sign" in the sense of "token".

An act of thought is an event and, therefore, defines a position in time. If our experiences always take place within the frame of *now* that means that every act of thought defines a point of reference. We cannot escape the *now* because the attempt to escape constitutes an act of thought and, therefore, defines a *now*. There is no thought without a point of reference because the thought itself defines it. This fact is expressed in grammar by the rule that every proposition must contain a verb, that is, a token-reflexive sign indicating the time of the event of which one speaks; for the tense of the verb has a token-reflexive meaning.

Token-reflexive meaning attaches also to the concepts *determined* and *undetermined*. The word "determination" denotes a relation between two situations A and B; the situation A determines, or does not determine, the situation B. It is meaningless to say that a situation B, considered separately, is determined. When we say that the past is determined, or that the future is undetermined, it is understood that one refers to the present situation; it is by reference to the *now* that the past is determined and the future is not. These words, as well as many others, have, therefore, token-reflexive meaning. Nevertheless it is true that these words express an objective relation; for it is a physical fact that if A is a situation defined by the act of speaking, a situation preceding A is determined with reference to A, whereas a situation succeeding A is not.

• *Index*

A CATALOG OF SELECTED
DOVER BOOKS
IN SCIENCE AND MATHEMATICS

A CATALOG OF SELECTED
DOVER BOOKS
IN SCIENCE AND MATHEMATICS

Astronomy

BURNHAM'S CELESTIAL HANDBOOK, Robert Burnham, Jr. Thorough guide to the stars beyond our solar system. Exhaustive treatment. Alphabetical by constellation: Andromeda to Cetus in Vol. 1; Chamaeleon to Orion in Vol. 2; and Pavo to Vulpecula in Vol. 3. Hundreds of illustrations. Index in Vol. 3. 2,000pp. 6⅛ x 9¼.
23567-X, 23568-8, 23673-0 Three-vol. set

THE EXTRATERRESTRIAL LIFE DEBATE, 1750–1900, Michael J. Crowe. First detailed, scholarly study in English of the many ideas that developed from 1750 to 1900 regarding the existence of intelligent extraterrestrial life. Examines ideas of Kant, Herschel, Voltaire, Percival Lowell, many other scientists and thinkers. 16 illustrations. 704pp. 5⅜ x 8½. 40675-X

A HISTORY OF ASTRONOMY, A. Pannekoek. Well-balanced, carefully reasoned study covers such topics as Ptolemaic theory, work of Copernicus, Kepler, Newton, Eddington's work on stars, much more. Illustrated. References. 521pp. 5⅜ x 8½.
65994-1

AMATEUR ASTRONOMER'S HANDBOOK, J. B. Sidgwick. Timeless, comprehensive coverage of telescopes, mirrors, lenses, mountings, telescope drives, micrometers, spectroscopes, more. 189 illustrations. 576pp. 5⅝ x 8¼. (Available in U.S. only.)
24034-7

STARS AND RELATIVITY, Ya. B. Zel'dovich and I. D. Novikov. Vol. 1 of *Relativistic Astrophysics* by famed Russian scientists. General relativity, properties of matter under astrophysical conditions, stars, and stellar systems. Deep physical insights, clear presentation. 1971 edition. References. 544pp. 5⅝ x 8¼. 69424-0

Chemistry

CHEMICAL MAGIC, Leonard A. Ford. Second Edition, Revised by E. Winston Grundmeier. Over 100 unusual stunts demonstrating cold fire, dust explosions, much more. Text explains scientific principles and stresses safety precautions. 128pp. 5⅜ x 8½. 67628-5

THE DEVELOPMENT OF MODERN CHEMISTRY, Aaron J. Ihde. Authoritative history of chemistry from ancient Greek theory to 20th-century innovation. Covers major chemists and their discoveries. 209 illustrations. 14 tables. Bibliographies. Indices. Appendices. 851pp. 5⅜ x 8½. 64235-6

CATALYSIS IN CHEMISTRY AND ENZYMOLOGY, William P. Jencks. Exceptionally clear coverage of mechanisms for catalysis, forces in aqueous solution, carbonyl- and acyl-group reactions, practical kinetics, more. 864pp. 5⅜ x 8½.
65460-5

THE HISTORICAL BACKGROUND OF CHEMISTRY, Henry M. Leicester. Evolution of ideas, not individual biography. Concentrates on formulation of a coherent set of chemical laws. 260pp. 5⅜ x 8½.   61053-5

A SHORT HISTORY OF CHEMISTRY, J. R. Partington. Classic exposition explores origins of chemistry, alchemy, early medical chemistry, nature of atmosphere, theory of valency, laws and structure of atomic theory, much more. 428pp. 5⅜ x 8½. (Available in U.S. only.)   65977-1

GENERAL CHEMISTRY, Linus Pauling. Revised 3rd edition of classic first-year text by Nobel laureate. Atomic and molecular structure, quantum mechanics, statistical mechanics, thermodynamics correlated with descriptive chemistry. Problems. 992pp. 5⅜ x 8½.   65622-5

Engineering

DE RE METALLICA, Georgius Agricola. The famous Hoover translation of greatest treatise on technological chemistry, engineering, geology, mining of early modern times (1556). All 289 original woodcuts. 638pp. 6¾ x 11.   60006-8

FUNDAMENTALS OF ASTRODYNAMICS, Roger Bate et al. Modern approach developed by U.S. Air Force Academy. Designed as a first course. Problems, exercises. Numerous illustrations. 455pp. 5⅜ x 8½.   60061-0

DYNAMICS OF FLUIDS IN POROUS MEDIA, Jacob Bear. For advanced students of ground water hydrology, soil mechanics and physics, drainage and irrigation engineering and more. 335 illustrations. Exercises, with answers. 784pp. 6⅛ x 9¼.
65675-6

ANALYTICAL MECHANICS OF GEARS, Earle Buckingham. Indispensable reference for modern gear manufacture covers conjugate gear-tooth action, gear-tooth profiles of various gears, many other topics. 263 figures. 102 tables. 546pp. 5⅜ x 8½.
65712-4

MECHANICS, J. P. Den Hartog. A classic introductory text or refresher. Hundreds of applications and design problems illuminate fundamentals of trusses, loaded beams and cables, etc. 334 answered problems. 462pp. 5⅜ x 8½.   60754-2

MECHANICAL VIBRATIONS, J. P. Den Hartog. Classic textbook offers lucid explanations and illustrative models, applying theories of vibrations to a variety of practical industrial engineering problems. Numerous figures. 233 problems, solutions. Appendix. Index. Preface. 436pp. 5⅜ x 8½.   64785-4

STRENGTH OF MATERIALS, J. P. Den Hartog. Full, clear treatment of basic material (tension, torsion, bending, etc.) plus advanced material on engineering methods, applications. 350 answered problems. 323pp. 5⅜ x 8½.   60755-0

A HISTORY OF MECHANICS, René Dugas. Monumental study of mechanical principles from antiquity to quantum mechanics. Contributions of ancient Greeks, Galileo, Leonardo, Kepler, Lagrange, many others. 671pp. 5⅜ x 8½.   65632-2

METAL FATIGUE, N. E. Frost, K. J. Marsh, and L. P. Pook. Definitive, clearly written, and well-illustrated volume addresses all aspects of the subject, from the historical development of understanding metal fatigue to vital concepts of the cyclic stress that causes a crack to grow. Includes 7 appendixes. 544pp. 5⅜ x 8½. 40927-9

STATISTICAL MECHANICS: Principles and Applications, Terrell L. Hill. Standard text covers fundamentals of statistical mechanics, applications to fluctuation theory, imperfect gases, distribution functions, more. 448pp. 5⅜ x 8½. 65390-0

THE VARIATIONAL PRINCIPLES OF MECHANICS, Cornelius Lanczos. Graduate level coverage of calculus of variations, equations of motion, relativistic mechanics, more. First inexpensive paperbound edition of classic treatise. Index. Bibliography. 418pp. 5⅜ x 8½. 65067-7

THE VARIOUS AND INGENIOUS MACHINES OF AGOSTINO RAMELLI: A Classic Sixteenth-Century Illustrated Treatise on Technology, Agostino Ramelli. One of the most widely known and copied works on machinery in the 16th century. 194 detailed plates of water pumps, grain mills, cranes, more. 608pp. 9 x 12. 28180-9

ORDINARY DIFFERENTIAL EQUATIONS AND STABILITY THEORY: An Introduction, David A. Sánchez. Brief, modern treatment. Linear equation, stability theory for autonomous and nonautonomous systems, etc. 164pp. 5⅜ x 8¼. 63828-6

ROTARY WING AERODYNAMICS, W. Z. Stepniewski. Clear, concise text covers aerodynamic phenomena of the rotor and offers guidelines for helicopter performance evaluation. Originally prepared for NASA. 537 figures. 640pp. 6⅛ x 9¼. 64647-5

INTRODUCTION TO SPACE DYNAMICS, William Tyrrell Thomson. Comprehensive, classic introduction to space-flight engineering for advanced undergraduate and graduate students. Includes vector algebra, kinematics, transformation of coordinates. Bibliography. Index. 352pp. 5⅜ x 8½. 65113-4

HISTORY OF STRENGTH OF MATERIALS, Stephen P. Timoshenko. Excellent historical survey of the strength of materials with many references to the theories of elasticity and structure. 245 figures. 452pp. 5⅜ x 8½. 61187-6

ANALYTICAL FRACTURE MECHANICS, David J. Unger. Self-contained text supplements standard fracture mechanics texts by focusing on analytical methods for determining crack-tip stress and strain fields. 336pp. 6⅛ x 9¼. 41737-9

Mathematics

HANDBOOK OF MATHEMATICAL FUNCTIONS WITH FORMULAS, GRAPHS, AND MATHEMATICAL TABLES, edited by Milton Abramowitz and Irene A. Stegun. Vast compendium: 29 sets of tables, some to as high as 20 places. 1,046pp. 8 x 10½. 61272-4

FUNCTIONAL ANALYSIS (Second Corrected Edition), George Bachman and Lawrence Narici. Excellent treatment of subject geared toward students with background in linear algebra, advanced calculus, physics and engineering. Text covers introduction to inner-product spaces, normed, metric spaces, and topological spaces; complete orthonormal sets, the Hahn-Banach Theorem and its consequences, and many other related subjects. 1966 ed. 544pp. 6⅛ x 9¼. 40251-7

ASYMPTOTIC EXPANSIONS OF INTEGRALS, Norman Bleistein & Richard A. Handelsman. Best introduction to important field with applications in a variety of scientific disciplines. New preface. Problems. Diagrams. Tables. Bibliography. Index. 448pp. 5⅜ x 8½. 65082-0

FAMOUS PROBLEMS OF GEOMETRY AND HOW TO SOLVE THEM, Benjamin Bold. Squaring the circle, trisecting the angle, duplicating the cube: learn their history, why they are impossible to solve, then solve them yourself. 128pp. 5⅜ x 8½. 24297-8

VECTOR AND TENSOR ANALYSIS WITH APPLICATIONS, A. I. Borisenko and I. E. Tarapov. Concise introduction. Worked-out problems, solutions, exercises. 257pp. 5⅝ x 8¼. 63833-2

THE ABSOLUTE DIFFERENTIAL CALCULUS (CALCULUS OF TENSORS), Tullio Levi-Civita. Great 20th-century mathematician's classic work on material necessary for mathematical grasp of theory of relativity. 452pp. 5⅝ x 8¼. 63401-9

AN INTRODUCTION TO ORDINARY DIFFERENTIAL EQUATIONS, Earl A. Coddington. A thorough and systematic first course in elementary differential equations for undergraduates in mathematics and science, with many exercises and problems (with answers). Index. 304pp. 5⅜ x 8½. 65942-9

FOURIER SERIES AND ORTHOGONAL FUNCTIONS, Harry F. Davis. An incisive text combining theory and practical example to introduce Fourier series, orthogonal functions and applications of the Fourier method to boundary-value problems. 570 exercises. Answers and notes. 416pp. 5⅜ x 8½. 65973-9

COMPUTABILITY AND UNSOLVABILITY, Martin Davis. Classic graduate-level introduction to theory of computability, usually referred to as theory of recurrent functions. New preface and appendix. 288pp. 5⅜ x 8½. 61471-9

ASYMPTOTIC METHODS IN ANALYSIS, N. G. de Bruijn. An inexpensive, comprehensive guide to asymptotic methods–the pioneering work that teaches by explaining worked examples in detail. Index. 224pp. 5⅜ x 8½ 64221-6

ESSAYS ON THE THEORY OF NUMBERS, Richard Dedekind. Two classic essays by great German mathematician: on the theory of irrational numbers; and on transfinite numbers and properties of natural numbers. 115pp. 5⅜ x 8½. 21010-3

APPLIED COMPLEX VARIABLES, John W. Dettman. Step-by-step coverage of fundamentals of analytic function theory—plus lucid exposition of five important applications: Potential Theory; Ordinary Differential Equations; Fourier Transforms; Laplace Transforms; Asymptotic Expansions. 66 figures. Exercises at chapter ends. 512pp. 5⅜ x 8½. 64670-X

INTRODUCTION TO LINEAR ALGEBRA AND DIFFERENTIAL EQUA-TIONS, John W. Dettman. Excellent text covers complex numbers, determinants, orthonormal bases, Laplace transforms, much more. Exercises with solutions. Undergraduate level. 416pp. 5⅜ x 8½. 65191-6

MATHEMATICAL METHODS IN PHYSICS AND ENGINEERING, John W. Dettman. Algebraically based approach to vectors, mapping, diffraction, other topics in applied math. Also generalized functions, analytic function theory, more. Exercises. 448pp. 5⅜ x 8¼. 65649-7

CALCULUS OF VARIATIONS WITH APPLICATIONS, George M. Ewing. Applications-oriented introduction to variational theory develops insight and pro-motes understanding of specialized books, research papers. Suitable for advanced undergraduate/graduate students as primary, supplementary text. 352pp. 5⅜ x 8½. 64856-7

COMPLEX VARIABLES, Francis J. Flanigan. Unusual approach, delaying com-plex algebra till harmonic functions have been analyzed from real variable view-point. Includes problems with answers. 364pp. 5⅜ x 8½. 61388-7

AN INTRODUCTION TO THE CALCULUS OF VARIATIONS, Charles Fox. Graduate-level text covers variations of an integral, isoperimetrical problems, least action, special relativity, approximations, more. References. 279pp. 5⅜ x 8½. 65499-0

CATASTROPHE THEORY FOR SCIENTISTS AND ENGINEERS, Robert Gilmore. Advanced-level treatment describes mathematics of theory grounded in the work of Poincaré, R. Thom, other mathematicians. Also important applications to problems in mathematics, physics, chemistry and engineering. 1981 edition. References. 28 tables. 397 black-and-white illustrations. xvii + 666pp. 6⅛ x 9¼. 67539-4

INTRODUCTION TO DIFFERENCE EQUATIONS, Samuel Goldberg. Excep-tionally clear exposition of important discipline with applications to sociology, psy-chology, economics. Many illustrative examples; over 250 problems. 260pp. 5⅜ x 8½. 65084-7

NUMERICAL METHODS FOR SCIENTISTS AND ENGINEERS, Richard Hamming. Classic text stresses frequency approach in coverage of algorithms, poly-nomial approximation, Fourier approximation, exponential approximation, other topics. Revised and enlarged 2nd edition. 721pp. 5⅜ x 8½. 65241-6

INTRODUCTION TO NUMERICAL ANALYSIS (2nd Edition), F. B. Hilde-brand. Classic, fundamental treatment covers computation, approximation, inter-polation, numerical differentiation and integration, other topics. 150 new problems. 669pp. 5⅜ x 8½. 65363-3

Physics

OPTICAL RESONANCE AND TWO-LEVEL ATOMS, L. Allen and J. H. Eberly. Clear, comprehensive introduction to basic principles behind all quantum optical resonance phenomena. 53 illustrations. Preface. Index. 256pp. 5⅜ x 8½.   65533-4

ULTRASONIC ABSORPTION: An Introduction to the Theory of Sound Absorption and Dispersion in Gases, Liquids and Solids, A. B. Bhatia. Standard reference in the field provides a clear, systematically organized introductory review of fundamental concepts for advanced graduate students, research workers. Numerous diagrams. Bibliography. 440pp. 5⅜ x 8½.   64917-2

QUANTUM THEORY, David Bohm. This advanced undergraduate-level text presents the quantum theory in terms of qualitative and imaginative concepts, followed by specific applications worked out in mathematical detail. Preface. Index. 655pp. 5⅜ x 8½.   65969-0

ATOMIC PHYSICS (8th edition), Max Born. Nobel laureate's lucid treatment of kinetic theory of gases, elementary particles, nuclear atom, wave-corpuscles, atomic structure and spectral lines, much more. Over 40 appendices, bibliography. 495pp. 5⅜ x 8½.   65984-4

AN INTRODUCTION TO HAMILTONIAN OPTICS, H. A. Buchdahl. Detailed account of the Hamiltonian treatment of aberration theory in geometrical optics. Many classes of optical systems defined in terms of the symmetries they possess. Problems with detailed solutions. 1970 edition. xv + 360pp. 5⅜ x 8½.   67597-1

THIRTY YEARS THAT SHOOK PHYSICS: The Story of Quantum Theory, George Gamow. Lucid, accessible introduction to influential theory of energy and matter. Careful explanations of Dirac's anti-particles, Bohr's model of the atom, much more. 12 plates. Numerous drawings. 240pp. 5⅜ x 8½.   24895-X

ELECTRONIC STRUCTURE AND THE PROPERTIES OF SOLIDS: The Physics of the Chemical Bond, Walter A. Harrison. Innovative text offers basic understanding of the electronic structure of covalent and ionic solids, simple metals, transition metals and their compounds. Problems. 1980 edition. 582pp. 6⅛ x 9¼.
66021-4

HYDRODYNAMIC AND HYDROMAGNETIC STABILITY, S. Chandrasekhar. Lucid examination of the Rayleigh-Benard problem; clear coverage of the theory of instabilities causing convection. 704pp. 5⅜ x 8¼.   64071-X

INVESTIGATIONS ON THE THEORY OF THE BROWNIAN MOVEMENT, Albert Einstein. Five papers (1905–8) investigating dynamics of Brownian motion and evolving elementary theory. Notes by R. Fürth. 122pp. 5⅜ x 8½.   60304-0

THE PHYSICS OF WAVES, William C. Elmore and Mark A. Heald. Unique overview of classical wave theory. Acoustics, optics, electromagnetic radiation, more. Ideal as classroom text or for self-study. Problems. 477pp. 5⅜ x 8½.   64926-1

PHYSICAL PRINCIPLES OF THE QUANTUM THEORY, Werner Heisenberg. Nobel Laureate discusses quantum theory, uncertainty, wave mechanics, work of Dirac, Schroedinger, Compton, Wilson, Einstein, etc. 184pp. 5⅜ x 8½.   60113-7

ATOMIC SPECTRA AND ATOMIC STRUCTURE, Gerhard Herzberg. One of best introductions; especially for specialist in other fields. Treatment is physical rather than mathematical. 80 illustrations. 257pp. 5⅜ x 8½.   60115-3

AN INTRODUCTION TO STATISTICAL THERMODYNAMICS, Terrell L. Hill. Excellent basic text offers wide-ranging coverage of quantum statistical mechanics, systems of interacting molecules, quantum statistics, more. 523pp. 5⅜ x 8½.   65242-4

THEORETICAL PHYSICS, Georg Joos, with Ira M. Freeman. Classic overview covers essential math, mechanics, electromagnetic theory, thermodynamics, quantum mechanics, nuclear physics, other topics. First paperback edition. xxiii + 885pp. 5⅜ x 8½.   65227-0

PROBLEMS AND SOLUTIONS IN QUANTUM CHEMISTRY AND PHYSICS, Charles S. Johnson, Jr. and Lee G. Pedersen. Unusually varied problems, detailed solutions in coverage of quantum mechanics, wave mechanics, angular momentum, molecular spectroscopy, more. 280 problems plus 139 supplementary exercises. 430pp. 6½ x 9¼.   65236-X

THEORETICAL SOLID STATE PHYSICS, Vol. 1: Perfect Lattices in Equilibrium; Vol. II: Non-Equilibrium and Disorder, William Jones and Norman H. March. Monumental reference work covers fundamental theory of equilibrium properties of perfect crystalline solids, non-equilibrium properties, defects and disordered systems. Appendices. Problems. Preface. Diagrams. Index. Bibliography. Total of 1,301pp. 5⅜ x 8½. Two volumes.   Vol. I: 65015-4   Vol. II: 65016-2

A TREATISE ON ELECTRICITY AND MAGNETISM, James Clerk Maxwell. Important foundation work of modern physics. Brings to final form Maxwell's theory of electromagnetism and rigorously derives his general equations of field theory. 1,084pp. 5⅜ x 8½. Two-vol. set.   Vol. I: 60636-8   Vol. II: 60637-6

OPTICKS, Sir Isaac Newton. Newton's own experiments with spectroscopy, colors, lenses, reflection, refraction, etc., in language the layman can follow. Foreword by Albert Einstein. 532pp. 5⅜ x 8½.   60205-2

THEORY OF ELECTROMAGNETIC WAVE PROPAGATION, Charles Herach Papas. Graduate-level study discusses the Maxwell field equations, radiation from wire antennas, the Doppler effect and more. xiii + 244pp. 5⅜ x 8½.   65678-5

INTRODUCTION TO QUANTUM MECHANICS With Applications to Chemistry, Linus Pauling & E. Bright Wilson, Jr. Classic undergraduate text by Nobel Prize winner applies quantum mechanics to chemical and physical problems. Numerous tables and figures enhance the text. Chapter bibliographies. Appendices. Index. 468pp. 5⅜ x 8½.   64871-0

METHODS OF THERMODYNAMICS, Howard Reiss. Outstanding text focuses on physical technique of thermodynamics, typical problem areas of understanding, and significance and use of thermodynamic potential. 1965 edition. 238pp. 5⅜ x 8½. 69445-3

TENSOR ANALYSIS FOR PHYSICISTS, J. A. Schouten. Concise exposition of the mathematical basis of tensor analysis, integrated with well-chosen physical examples of the theory. Exercises. Index. Bibliography. 289pp. 5⅜ x 8½. 65582-2

RELATIVITY IN ILLUSTRATIONS, Jacob T. Schwartz. Clear nontechnical treatment makes relativity more accessible than ever before. Over 60 drawings illustrate concepts more clearly than text alone. Only high school geometry needed. Bibliography. 128pp. 6⅛ x 9¼. 25965-X

THE ELECTROMAGNETIC FIELD, Albert Shadowitz. Comprehensive undergraduate text covers basics of electric and magnetic fields, builds up to electromagnetic theory. Also related topics, including relativity. Over 900 problems. 768pp. 5⅝ x 8¼. 65660-8

GREAT EXPERIMENTS IN PHYSICS: Firsthand Accounts from Galileo to Einstein, edited by Morris H. Shamos. 25 crucial discoveries: Newton's laws of motion, Chadwick's study of the neutron, Hertz on electromagnetic waves, more. Original accounts clearly annotated. 370pp. 5⅜ x 8½. 25346-5

RELATIVITY, THERMODYNAMICS AND COSMOLOGY, Richard C. Tolman. Landmark study extends thermodynamics to special, general relativity; also applications of relativistic mechanics, thermodynamics to cosmological models. 501pp. 5⅜ x 8½. 65383-8

LIGHT SCATTERING BY SMALL PARTICLES, H. C. van de Hulst. Comprehensive treatment including full range of useful approximation methods for researchers in chemistry, meteorology and astronomy. 44 illustrations. 470pp. 5⅜ x 8½. 64228-3

STATISTICAL PHYSICS, Gregory H. Wannier. Classic text combines thermodynamics, statistical mechanics and kinetic theory in one unified presentation of thermal physics. Problems with solutions. Bibliography. 532pp. 5⅜ x 8½. 65401-X

Paperbound unless otherwise indicated. Available at your book dealer, online at **www.doverpublications.com**, or by writing to Dept. GI, Dover Publications, Inc., 31 East 2nd Street, Mineola, NY 11501. For current price information or for free catalogues (please indicate field of interest), write to Dover Publications or log on to **www.doverpublications.com** and see every Dover book in print. Dover publishes more than 500 books each year on science, elementary and advanced mathematics, biology, music, art, literary history, social sciences, and other areas.